T0353401

QUANTUM THEORY OF THE CHEMICAL BOND

QUANTUM THEORY OF THE CHEMICAL BOND

by

R. DAUDEL

Sorbonne and Center of Applied Wave Mechanics (C.N.R.S.), Paris, France

D. REIDEL PUBLISHING COMPANY

DORDRECHT-HOLLAND / BOSTON-U.S.A.

THÉORIE QUANTIQUE DE LA LIAISON CHIMIQUE
First published in 1971 by the Presses Universitaires de France
Translated by Nepean Translation Service

Library of Congress Catalog Card Number 74–82700

Cloth edition: ISBN 90 277 0264 0
Paperback edition: ISBN 90 277 0528 3

Published by D. Reidel Publishing Company,
P.O. Box 17, Dordrecht, Holland

Sold and distributed in the U.S.A., Canada, and Mexico
by D. Reidel Publishing Company, Inc.
306 Dartmouth Street, Boston,
Mass. 02116, U.S.A.

Printed in The Netherlands by D. Reidel, Dordrecht

TABLE OF CONTENTS

INTRODUCTION

The present text is a rational analysis of the concept of the chemical bond by means of the principles of wave mechanics. The discussion of the material has been arranged so as to render its main content comprehensible for readers who may not have had previous training in quantum mechanics.

The text comprises three major parts. It begins with an exposition of the fundamental ideas. In this section the principles are reviewed from which de Broglie developed his mechanics; this allows the book to be read by chemistry majors and freshmen alike. However, we believe that it may also be of interest to university- and college teachers who must include certain aspects of quantum chemistry into their courses while being insufficiently familiar with the subject. It may even be of interest to science teachers in secondary schools. Finally, having been a witness to the evolution of these notions for over a quarter of a century, we present certain concepts from a particular point of view which might prove attractive to chemists of all kinds, perhaps even quantum chemists.

The second, more technical part summarizes the methods of constructing wave functions that describe the electrons in molecules. This section can only be fully appreciated by those readers who are familiar with some aspects of the algorithms used in quantum mechanics.

On the other hand, we have expended considerable effort on writing the third part of this book in such a way that reference is only occasionally made to the techniques described in the second part. Hence it should be possible to study part three without having read part two. Some applications of the quantum theory of the chemical bond are presented in this last section.

In order to stress the multitude of areas where this theory demonstrates its value, we shall treat problems of reaction mechanisms and subjects of interest to biochemistry, biology, pharmacology as well as questions relevant to industrial chemistry.

CHAPTER I

FUNDAMENTAL IDEAS

1. Introduction

Only three years separate the birth of wave mechanics [1]* from its application to the study of the chemical bond [2a, b]. I shall therefore try to report here some of the principal results obtained in the course of the subsequent forty years, during which the methods of mathematical physics penetrated into the field of chemistry. In our opinion it is useful first of all to place certain epistemological aspects of this amalgamation in perspective.

As in every other domain of mathematical physics, one stands in witness of the origin of a notion shaped by the intimate contact between an applied rationalism and a technical materialism [3]. The dialogue between the theoretician and the experimenter who in the initial stages of his research often operates on a qualitative, intuitive level, is carried on until an ensemble of mathematical concepts has been brought to bear on an ensemble of notions derived from the experiment though appearing in a quantitative structure. We shall examine in more detail how such a connection is established.

Concepts in physics and chemistry alike tend to become increasingly *operational* in the sense of Bridgman [4].

In order to find the length of an object we have to execute certain physical operations. The concept of length is then established when the operations measuring length have been fixed, i.e. the concept of length is not more and not less than the ensemble of operations measuring the length. In general, a concept is nothing more than an ensemble of operations: the concept is synonymous with its corresponding operations.

Is it not precisely so with the length of a chemical bond? Which chemist can pretend to have 'seen' such a length? And, strictly speaking, how is a number ultimately determined if not by the set of operations eventually leading to this number? We intend to show that such a number which is situated at the point where applied rationalism and technical materialism flow together, acquires the dualism of simultaneously being true and real. Its reality stems from the engendering material operations practised by the physical chemist, and the multitude

* Numbers in brackets refer to the list of references on p. 105.

of paths converging on the same number. It is as in the case that two series of operations as different as electron diffraction and microwave spectroscopy lead to the same number within experimental error. Then its reality stems from the fact that when one solves the wave equation associated with the molecule in question, the same number is found again. Here one applies a sequence of purely rational operations and relies on pencil and paper only to support one's short-term memory. One cannot emphasize often enough the miracle of such concurrent findings which give the scientist the feeling of security so characteristic of the mature scientific method. Operationalism induces the real and the true to bring forth the same entity: the number. Pythagoras must have had a presentiment when he noticed a numerical and harmonic duality in reality; and, in fact, the number remains the only indisputable point where applied rationalism and technical materialism flow together. The theory is allright by definition when for the calculation of a quantity it leads to a sequence of operations providing the same number as the experimental manipulations eventually measuring it. Theory and experiment derive their very affinity from the efforts made by the theoreticians and the experimenters in their attempts to arrive at the same numerical values. When the theoretical concepts are given careful scrutiny, it will soon become evident that they are of a very distinct kind. One should bear in mind that the first step in the sequence of operations creating the concepts, is to associate an entire molecule with an equation, the wave equation. The principles of wave mechanics provide the precise rules according to which such association can be conveniently carried out. The solution of the wave equation is a function, the total wave function, which enables us to calculate all those characteristic quantities that otherwise cannot be determined. The fundamental concepts which form the subject matter of the first part of this book will be grouped around the total wave function with its dependence on all the electrons of the molecule. Unfortunately, there are no practicable means to find an exact wave function, except in the case of extremely simple molecules. For this reason the quantum theory of the chemical bond utilizes numerous approximation methods. Moreover, there is the remarkable fact that some concepts are defined only by means of approximate wave functions, and gradually vanish the more exact the wave function becomes. For example, the charge distribution in a molecule, bond orders and free valencies are such concepts. It is evident that such concepts have a technical rather than a physical flavour, that they are based on cognition rather than on physical reality; they will be encountered in the second part of this book.

Among the procedures for the construction of wave functions the most common one is that which develops the function from a basis of orbitals. It will be shown that this method is largely analogous to the representation of a vector by its components in a coordinate system. It is possible to rotate the basis of the orbitals leaving the approximate wave function invariant, though changing its expression in terms of the orbitals. This is similar to a rotation of the coordinate axes whereby the components of the vector are alterated. Strangely enough, some concepts have been defined so as to be dependent on the choice of the orbital basis of the wave function. The exchange energy belongs to this class. One senses that such concepts are bound to be physically meaningless; they can never coincide with measurable quantities, and populate physical parlance with dangerous expressions suggesting many false notions to the minds of those not yet immunized by a period of prolonged asceticism. We will call attention to such ambiguous concepts as soon as they appear, since it is just as intolerable not to give a feeling of the physical meaning behind a concept as it is to make others believe that there is a meaning hidden behind a concept actually devoid of such meaning.

2. Wave Mechanics

Differences are worthwhile since they create the most beautiful harmony, for everything is shaped by struggle. (...) They do not understand how the differences are reconciled; the harmony of opposite tensions like those of the arch and the lyra.

Thus wrote Heraclitus of Ephesus five centuries before our era. Wave mechanics gives a view of the world with the discontinuous and the continuous living in harmony, like the particle and the wave, the certain and the uncertain. For the benefit of readers untrained in such usage, the principal ideas upon which the present theoretical structure is based, will be brought together.

THE DE BROGLIE WAVE AND THE PARTICLE

A study of light will very quickly reveal its dual nature, continuous and discontinuous. In the geometrical theory of Descartes, which associates each bundle of light with a collection of straight lines, rays, it is considered as continuous. Isaac Newton already considered each brilliant object as a generator of light particles, grains with a very small weight and forcefully ejected into space. Applying the principles of his mechanics to this discontinuous phenomenon, Newton gave an elegant interpretation of reflection and refraction. The discovery of interference fringes and diffraction effects forced Fresnel to treat light as though it were a transverse wave propagating along the rays.

Only this structure, continuous again, permitted him to understand why a doubly illuminated area may appear dark. Finally the interpretation of the experimental data on the photo-electric effect led Einstein to the rehabilitation of the light corpuscle while preserving its wave character. In this theory a monochromatic radiation appears as a kind of undulatory fluid characterized by a certain frequency v with its energy concentrated in particles called photons.

In the meantime, Planck was led to assume that the exchange of energy between radiation and matter could take place in small, distinct quantities only, i.e. in quanta W proportional to the frequency v of the wave-like intermedium. Such a hypothesis permitted a theoretical derivation of the distribution of the radiation frequencies emitted by a black body. The factor h, connecting W and v, bears Planck's name. As a natural consequence of this Einstein supposed that the energy of a photon guided by a wave, should be expressed by

$$W = hv.$$

This relation enabled him to give a quantitative explanation of the fundamental laws of the photo-electric effect. Fresnel's wave-theory associates every light ray travelling along the x-axis, with a wave

$$\Psi(x, t) = a \cos 2\pi [vt - x/\lambda],$$

where λ = wavelength.

According to Fresnel, this is equivalent to the complex form

$$\Psi(x, t) = a \exp[2\pi i(vt - x/\lambda)].$$

Furthermore, light intensity is proportional to the square of the amplitude a^2, i.e. to the square of the modulus of Ψ. According to Einstein's hypothesis which reduces the energy to a collection of identical corpuscles, the light intensity must be proportional to the volume density of those corpuscles. In view of this the volume density of the photons must be proportional to a^2 or to $|\Psi|^2$.

In 1923 Louis de Broglie became convinced that the wave-particle dualism is not an exclusive attribute of the photon, but rather that every particle can be associated with a wave. He clearly anticipated the wave character of electricity. Davisson and Germer's experiments in 1927 proved the correctness of that idea when these scientists proved that an electron beam produces diffraction fringes just like a light beam does. In 1924 Louis de Broglie had already defined distinct links between waves and particles. In particular, he showed that for the classical trajectory of a particle with mass m and velocity v to coincide with an associated wave within the limits of the approximations of geometrical optics, the wavelength should be selected so as to obey the relation

$$\lambda = h/mv,$$

which is de Broglie's formula. The wavelength calculated with this formula for the electrons in Davisson and Germer's experiments, accounted quantitatively for the structure of the observed diffraction pattern. So far we have been interested in light rays associated with waves, and hence in large numbers of photons. When we have the intention to study the chemical bond, the wave function must describe very small numbers of photons. We must therefore scrutinize the significance of an associated wave relative to decreasing numbers of particles associated with it. We noted that for large numbers the square of the modulus of the wave, $|\Psi(M, t)|^2$, represents the volume density of the particles at point M at time t. In the extreme case of a wave associated with one electron only, it is obvious that $|\Psi(M, t)|^2$ can no longer represent the number of electrons in one unit volume at point M, since this number would be zero everywhere except at the very point where the electron happens to be. Brief reflection yields the recognition of a probability as the only possible link between wave and particle. Let us therefore assume that the probability dp of finding the electron in a volume dv around point M at time t, shall be given by

$$dp = |\Psi(M, t)|^2 \, dv.$$

This means that

$$dp/dv = |\Psi(M, t)|^2$$

represents a probability density; this becomes an ordinary density when the number of particles becomes large, and by virtue of the laws of large numbers these probabilities will turn into certainties.

THE WAVE EQUATION

It can easily be seen that the wave travelling along the x-axis,

$$\Psi = a \exp[2\pi i (vt - x/\lambda)]$$

obeys the equation

$$d^2\Psi/dx^2 = v^{-2}\lambda^{-2} \, d^2\Psi/dt^2$$

In the general context of classical wave optics it can be shown that a light wave

$$\Psi(M, t) = a(M) \exp[2\pi i (vt - \varphi(M))]$$

obeys d'Alembert's equation

$$\partial^2\Psi/\partial x^2 + \partial^2\Psi/\partial y^2 + \partial^2\Psi/\partial z^2 = v^{-2}\lambda^{-2}\partial^2\Psi/\partial t^2$$

which can be written in a more concise form as*

$$\nabla\Psi = v^{-2}\lambda^{-2}\partial^2\Psi/\partial t^2.$$

---

* The symbol ∇ (nabla) identifies the Laplace operator, defined as the summation of second derivatives:

$$\frac{\partial^2}{\partial x^2} + \frac{\partial^2}{\partial y^2} + \frac{\partial^2}{\partial z^2}$$

(for simplicity's sake ∇ is used instead of the customary ∇^2).

We note that

$$\frac{\partial \Psi}{\partial t} = 2\pi i v \Psi; \qquad \frac{\partial^2 \Psi}{\partial t^2} = -4\pi^2 v^2 \Psi.$$

Substitution of the latter expression into d'Alembert's equation yields

$$\nabla \Psi = -\left(4\pi^2/\lambda^2\right)\Psi. \tag{1.1}$$

Suppose now that the wave Ψ is associated with an electron and that its total energy E can be separated into a potential energy $F(M)$ and a kinetic energy $mv^2/2$. Then

$$E = F(M) + \tfrac{1}{2}mv^2.$$

But since

$$\lambda = h/mv$$

we may write

$$E = F(M) + h^2/(2m\lambda^2)$$

and so

$$\lambda^{-2} = \left(2m/h^2\right)\left(E - F(M)\right)$$

which then is substituted into (1):

$$\nabla \Psi = -\left(8\pi^2 m/h^2\right)\left(E - F(M)\right)\Psi$$

or

$$-\left(h^2/8\pi^2 m\right)\nabla \Psi + F(M)\,\Psi = E\Psi.$$

This is the so-called Schrödinger equation. Its solution gives the exact form of the function $\Psi(M, t)$ representing the distribution of the wave over space and time, thus permitting us to calculate at any moment the probability of finding the particle in any selected domain of space.

STATIONARY STATES

Let us briefly review the fundamental ideas encountered so far. We shall associate a wave of frequency v with a particle of energy E, such that

$$E = hv.$$

At point M and time t this may be written

$$\Psi(M, t) = a(M)\exp 2\pi i v t.$$

(One should notice that the quantity $\exp 2\pi i \varphi(M)$ can always be absorbed into $a(M)$.)

When this wave describes a particle with mass m and subject to such interactions as to give it a potential energy $F(M)$, the wave obeys the equation

$$-(h^2/8\pi^2m)\,\nabla\Psi + F(M)\,\Psi = E\Psi.$$

The probability dp of finding the particle in a volume dv at time t is given by the expression

$$dp = |\Psi\,(M,\,t)|^2\,dv.$$

This is why the derivative

$$dp/dv = |\Psi\,(M,\,t)|^2$$

is called the probability density. Essentially, the wave appears to have the nature of a probability; it is a probability wave. Note that, according to the well-known properties of exponential functions

$$|\Psi\,(M,\,t)|^2 = |a\,(M)|^2.$$

Consequently, the probability of finding the particle in the volume element dv is independent of time. Hence, the wave in question describes a stable state, a state not changing with time: it is a *stationary state*. Notice also that since the particle cannot fail to be found in space somewhere, the summation of the probabilities dp corresponding with all the volume elements dv of that space must be the same as a certainty, i.e. must equal 100% or 1. This means that,

$$\int_{\text{space}} dp = \int_{\text{space}} |\Psi\,(M,\,t)|^2\,dv = 1.$$

The wave function $\Psi\,(M,\,t)$ is normalized to unity. Next we wish to point out that with

$$H = -(h^2/8\pi^2m)\,\nabla + F$$

the Schrödinger equation becomes

$$H\Psi = E\Psi,$$

which clearly shows that the operations performed by the operator H on the function Ψ must lead to a simple multiplication of the function by a constant, the energy E of the particle. The function Ψ is called the *eigenfunction* of the operator H and the energy E is the corresponding *eigenvalue*. These ideas will now be applied to the movements of a particle with mass m and confined to a segment of a straight line with length l (Figure 1).

Fig. 1.

The x-axis is chosen collinear with the segment, one end of which is in the origin. In order to prevent the particle from leaving the segment, it is subjected to a potential which is constant along the segment and infinitely repulsive beyond its end points. The wave equation is then simply

$$-(h^2/8\pi^2\,m)\,\mathrm{d}^2\Psi/\mathrm{d}x^2 + F(x)\,\Psi = E\Psi$$

with $F(x)$=Constant for all x on the segment, but infinite for all x outside of the segment. This equation will now be solved for the open interval $(0, l)$. Within this interval the potential energy is defined apart from a constant which can always be taken so as to obey

$F(x)=0$ for all x in the open interval $(0, l)$.

Within the interval we have

$$-(h^2/8\pi^2 m)\,\mathrm{d}^2\Psi/\mathrm{d}x^2 = E\Psi,$$

where
$$\Psi = a(x)\exp 2\pi i v t.$$

It follows quite easily that

$$-(h^2/8\pi^2 m)\,(\mathrm{d}^2 a/\mathrm{d}x^2) = Ea \qquad (1.2)$$

and from this

$$a = k\sin(Ax + B). \qquad (1.3)$$

The calculation of A and B is easy:

$$(\mathrm{d}^2 a/\mathrm{d}x^2) = -A^2 a$$

which after substitution into (1.2) leads to

$$(h^2 A^2/8\pi^2 m) = E.$$

Now the equation's behaviour must be examined for the case that $x \to 0$. Write

$$-(h^2/8\pi^2 m)\,(\mathrm{d}^2 a/\mathrm{d}x^2) + aF(x) = Ea.$$

The right hand side, Ea, remains finite because a is a continuous function. Consequently, the left hand side must also remain finite. The first term fulfills that requirement. The second term must behave likewise. Since $F(x) \to \infty$ it is necessary that $a \to 0$. And so for $x=0$

$$a = k\sin(Ax + B) = 0.$$

Hence
$$\sin B = 0; \qquad B = p\pi.$$

So we may let $B=0$ without imposing restrictions on the generality of the discussion, and the function becomes

$$a = k \sin Ax \quad \text{with} \quad (h^2 A^2 / 8\pi^2 m) = E. \tag{1.4}$$

The same kind of reasoning shows that when $x \to l$ it is necessary that $a \to 0$, and consequently

$$O = k \sin Al; \quad Al = n\pi \quad \text{with} \quad n = \text{integer.}$$

Substitution of this value of A into (1.4) eventually yields

$$E = (h^2 n^2 / 8ml^2) \tag{1.5}$$

and

$$a = k \sin (n\pi / l) \, x. \tag{1.6}$$

This result indicates that the energy of the particle can only assume one of the values given by formula (1.5). It cannot vary continuously. It undergoes discontinuous changes only, passing from one value of the set defined by (1.5) to another value in the same set. The energy is said to be *quantized* and dependent on the *quantum number n*.

It will be clear, however, that the wave's amplitude is a sinusoidal function of x. So it equals zero for a series of values x, and passes through maxima for other values x. The wave possesses *nodes* and *antinodes*. This resembles the classical stationary waves (e.g. Melde's experiment, resonance of sound waves in a tube). For $n = 1$ there is one node at each end of the segment and a maximum at the center. The best chances to find the particle are in the center. It will never go to the very ends since that is forbidden. For $n = 2$ a third node appears in the middle of the segment; the particle is no longer allowed to be there. The maximum of the probability density corresponds with

$$x = \tfrac{1}{4} l \quad \text{and} \quad x = \tfrac{3}{4} l.$$

One observes that with increasing n, i.e. with increasing energy, the number of nodes increases. The distribution of the probability density depends on the energy.

WAVE MECHANICS AS A SET OF THEOREMS

The essentials of wave mechanics can be condensed to three principles.

(a) *The Principle of the Associated Wave*

Each particle with mass m and subjected to a force field giving it a potential energy $F(M)$ at point M, is associated with a wave (M, t). This wave is a solution of

$$- (h^2 / 8\pi^2 m) \, \nabla \Psi + F(M) \, \Psi = (h/2\pi i) \, \partial \Psi / \partial t. \tag{1.7}$$

The square of the modulus of this wave represents the probability density of the particle at point M and time t. This equation can be recognized as a generalization of the Schrödinger equation. It can be applied to a system in evolution.

In the case of a stationary state where

$$\Psi = a(M) \exp 2\pi i v t$$

one has

$$\partial \Psi / \partial t = 2\pi i v \Psi = (2\pi i / h) E \Psi,$$

and so

$$-(h^2/8\pi^2 m) \nabla \Psi + F(M) \Psi = E \Psi.$$

This is Schrödinger's equation.

(b) *The Principle of Quantization*

Each quantity is associated with a *linear* and *Hermitean operator*. Every exact value of that quantity is an eigenvalue of the operator. We have already associated the energy with an operator

$$H = -(h^2/8\pi^2 m) \nabla + F$$

and found that for a stationary state it could only assume one of the eigenvalues of H.

(c) *The Principle of the Spectral Decomposition*

Be A the operator associated with a certain quantity, and be f_i its ith eigenfunction. When the wave function is expanded into the series

$$\Psi(M, t) = \sum_i c_i f_i$$

the probability that a measured value of that quantity coincides with the ith eigenvalue of operator A, equals $|c_i|^2$.

3. One-Electron Systems; the Hydrogen Atom and -Molecular Ion

THE HYDROGEN ATOM ACCORDING TO BOHR'S THEORY

It is very instructive to juxtapose the pictures of the hydrogen atom provided by respectively the old quantum mechanics and wave mechanics. These intellectual concepts born of the same model and rich in contradictions and analogies, shed a clear light on the connections and divergences of the theses that brought them forth. Although undoubtedly all readers are familiar with Bohr's theory, we found it desirable even so to group together here its principal elements.

Let us return to its origin. Early in this century a choice had to be made between two models of the atom. The experiments by Geiger and Marsden and their interpretation by Rutherford imposed the choice of Jean Perrin's planetary model. This led the physicists into a dead alley. In that model the electrons circle around a positively charged nucleus. Because of this they are in a state of acceleration and, when one has faith in the laws of classical electromagnetism, such electrons must emit energy continuously and eventually fall onto the nucleus. Jean Perrin's model is therefore proved to be fundamentally unstable in the context of a classical theory. Fortunately, a new mechanics called quantummechanics was conceived on the foundations laid by Planck's work. We did already mention Planck's hypothesis which, for the explanation of the black-body radiation, had to assume that exchange of energy between matter and radiation could not take place but with energy quanta h. The mathematical analysis of this proposition led to the more general principle expressed by the statement that the momentum over one period of a mass m executing a periodical motion, necessarily equals an integer multiple of Planck's constant h. Applying this principle Bohr could without difficulty stabilize Perrin's model. In his model the single electron of a hydrogen atom, with a charge $-e$ and a mass m, moves around the nucleus with charge $+e$, a proton, in a circular orbit with radius r under the influence of a force given by Coulomb's law

$$F = e^2/r^2. \tag{1.8}$$

With γ for its acceleration, one finds

$$F = m\gamma = mv^2/r \tag{1.9}$$

and by combining (1.8) and (1.9)

$$mv^2 = e^2/r. \tag{1.10}$$

For one circular orbit one finds then

$$2\pi rmv = nh; \qquad n = \text{integer} \tag{1.11}$$

from which follows

$$4\pi^2 r^2 m^2 v^2 = n^2 h^2. \tag{1.12}$$

Introducing into (1.12) the value of mv^2 given by (1.10), one obtains

$$r = (n^2 h^2)/(4\pi^2 m e^2) \tag{1.13}$$

The orbits of the electrons are therefore not arbitrary circles. The radius of any circle must obey formula (1.13) with $n=$integer. The radius of an orbit cannot change continuously but changes stepwise with changing n: it is quantized. The total energy of an electron orbiting in such a circle will now be calculated. It suffices to add the potential energy $-e^2/r$ to the kinetic energy $\frac{1}{2}mv^2$

$$E = \tfrac{1}{2}mv^2 - e^2/r. \tag{1.14}$$

Taking (1.10) into consideration, this becomes

$$E = -\tfrac{1}{2}e^2/r \tag{1.15}$$

or rather, expressing r through formula (1.13)

$$E_n = - (2\pi^2 me^4)/(n^2 h^2). \tag{1.16}$$

As long as the electron remains in a given orbit its energy remains the same. So we must assume that contrary to the prediction of classical electromagnetics, it does not radiate. Bohr assumed that an atom can only emit or absorb radiation when one of its electrons changes orbits, and that then it emits or absorbs a photon with an energy $W = hv$. The principle of conservation of energy leads us to

$$W = hv = E_i - E_f = (2\pi^2 me^4/h^2)\,(1/f^2 - 1/i^2), \tag{1.17}$$

where i and f denote the values of the quantum numbers n associated with respectively the initial and final orbits involved in the transition phenomenon.

We know that with this formula we can calculate the exact frequencies of the radiation leaving a tube containing hydrogen and through which an electric discharge is passed.

It is customary to say that an electron circling in the nth orbit is in the nth shell of the atom. These shells are referred to as K-, L-, M-, N-, O-, P-, Q-, ... shell for $n = 1, 2, 3, 4, 5, 6, 7$ respectively. According to this picture an emission or absorption of light by an atom can be associated with an electron moving from one shell to another. Finally we note that formula (1.13) assigns a radius of 0.53 Å to the orbit belonging to the K-shell.

THE HYDROGEN ATOM IN WAVE MECHANICS

We shall retain the same planetary model of the hydrogen atom but will treat it now according to the principles of wave mechanics. As a matter of simplification we replace the proton by a charge $+e$ in a fixed place in space. This simplification can be shown to be very convenient. The electron shall have an associated wave function $\Psi(M, t)$, the solution of the equation

$$- (h^2/8\pi^2 m)\, \nabla\Psi + F(M)\,\Psi = (h/2\pi i)\, \partial\Psi/\partial t.$$

In the point M the electron experiences a potential energy

$$F(M) = - e^2/r,$$

where r is the distance between M and the nucleus. In this case, however, one should consider the equivalent of an electron steadily circling in an orbit, as a stationary state. This reduces the wave equation to

$$- (h^2/8\pi^2 m) \nabla \Psi - (e^2/r) \Psi = E\Psi \qquad (1.18)$$

with

$$\int_{space} |\Psi|^2 \, dv = 1. \qquad (1.19)$$

Also, a stable atom corresponds with a negative energy (attraction). It can be shown that Equation (1.18) has a solution satisfying condition (1.19) only when

$$E = - (2\pi^2 m e^4)/(n^2 h^2),$$

where n=integer. When a hydrogen atom is in a stationary state, the energy of its electron will be quantized. Remarkably enough, it is given by a formula which turns out to be the same as the one implied in Bohr's theory. Thus the frequencies of the radiation emitted by a hydrogen atom are found again, this time by assuming that an electron absorbs or emits radiation only when passing from one stationary state to another. One may continue to speak the same language and say that the electron is in a shell n when its energy is given by

$$E_n = - (2\pi^2 m e^4)/(n^2 h^2).$$

It is not necessary to infer as in Bohr's theory, that the electron moves in a circular orbit; but what kind of picture does it then provide of the hydrogen atom? This question will be answered for the electron in the K-shell, when its energy is given by

$$E_1 = - (2\pi^2 m e^4/h^2).$$

One notices that the solution of the wave equation in this case is simply

$$\Psi (M, t) = k \exp - (r/a_0) \exp (2\pi i E_1 t/h)$$

when one takes

$$a_0 = h^2/4\pi m e^2,$$

a value coinciding with the radius of the K-shell in Bohr's theory. The electron's probability density in point M at distance r from the nucleus may then be written as

$$|\Psi (M, t)|^2 = k^2 \exp - (2r/a_0).$$

Figure 2 is a symbolic representation of this function. Apparently of all the points in space the one most frequently visited by the electron is the proton, the nucleus of the hydrogen atom.

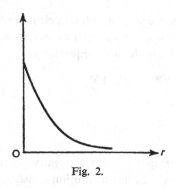

Fig. 2.

The ancient and modern descriptions of the hydrogen atom, although identical with regard to energy, do not lead to the same geometrical picture. In Bohr's theory the electron remains scrupulously at a good distance from the nucleus (0.53 Å or 0.53×10^{-8} cm), in wave mechanics it skirts along its surface, from time to time.

Nuclear physics suggests, within a radius of 10^{-13} cm around a proton, the presence of a specific field capable of transforming an electron into a neutral particle, a neutrino, whereby the proton becomes a neutron:

$$p + e \rightarrow n + v; \qquad v = \text{neutrino}.$$

In the case of hydrogen this reaction is impossible for reasons of energy. It would be reasonable, though, to keep this possibility in mind for other atoms. In 1935, Yukawa and Sakata [5] expressed their belief in the capability of certain atomic nuclei to absorb spontaneously one of their electrons, thereby undergoing a transmutation into the element belonging to the position in the periodic system preceding its original one. The discovery by Alvarez in 1937 [6] of the phenomenon of electron capture provided a brilliant confirmation of this hypothesis, and at the same time it proved that the electrons of an atom can penetrate to within a distance of the order of 10^{-13} cm from the nucleus, i.e. 10^5 times closer than is allowed by the smallest Bohr radius. This result indicated the practical superiority of wave mechanics.

It can be shown that for an arbitrary shell n the wave function $\Psi(M, t)$ associated with the electron in a hydrogen atom, may be written

$$\Psi(M, t) = R_{n,l}(r)\, \Theta_{l,m}(\theta)\, \Phi_m(\varphi) \exp(2\pi i E_n t / h), \qquad (1.20)$$

where r, θ and φ refer to point M as indicated in Figure 3. The radial function $R_{n,l}(r)$ is dependent on the integers n and l. Earlier we defined the quantum number n; l is another quantum number which may assume any value 0 to $(n-1)$ for any given n. When an electron is characterized by $l = 0, 1, 2, 3, \ldots$ it is said to be in a s-, p-, d-, f-, $\ldots$ state. This gives n possible states for an electron in

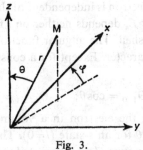

Fig. 3.

shell n. Figure 4 presents the graphs of several functions R. With increasing n the wave function is seen to become increasingly important at larger values of r. The electron tends to move away from the nucleus. For an s-state $(l=1)$ the function attains a maximum at $r=0$; on the other hand, it equals zero for any other state. The function $\Phi_m(\varphi)$ depends on a third quantum number, m. Being integer or zero, it is in the closed interval $(-l, +l)$ for a state with quantum number l; there are, therefore, $(2l+1)$ possible values of m. An electron is said to have σ-, π-, δ-, φ-,... character for the respective values of $l=0, \pm 1, \pm 2, \pm 3, \ldots$.

Equation (1.20) shows that

$$|\Psi(M, t)|^2 = |R_{n,l}(r)|^2 |\Theta_{l,m}(\theta)|^2 |\Phi_m(\varphi)|^2.$$

This is no more than a schematic representation giving the general shape of the curves, devoid of any quantitative merit.

It can be shown that, apart from a constant,

$$\Phi_m(\varphi) = \exp im\varphi.$$

It suffices then to write

$$|\Psi(M, t)|^2 = |R_{n,l}(r)|^2 |\Theta_{l,m}(\theta)|^2.$$

In a stationary state corresponding with well-defined values of n, l and m the

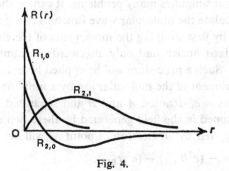

Fig. 4.

probability density of the electron is independent of the angle φ. In addition one may notice that the energy E_n depends neither on l nor on m. The energy is a characteristic value of the shell. The angular function $\Theta_{l,m}$ depends on both l and m, i.e. on state and character. Except for a constant that can be absorbed into $R_{n,l}(r)$, we have

$$\Theta_{0,0} = 1; \qquad \Theta_{1,0} = \cos\theta; \qquad \Theta_{1,\pm1} = \sin\theta.$$

The probability of finding the electron in a volume element dv around the point M is independent of θ for an s-state ($l=0$). The density's center of symmetry is the nucleus. For a p-state ($l=1$) the distribution of the density depends on the character. When the character is established by $m=0$, it varies as $\cos^2\theta$ for one particular value of r. It has a maximum along the z-axis which is an axis of symmetry of the density distribution. In the case of π-character ($m=\pm1$) the density varies as $\sin^2\theta$. It reaches a maximum in the $0xy$-plane which acts as a symmetry plane. Finally, notice that the probability of finding the electron within a sphere with radius $r=1.67$ Å centered around the nucleus, equals 92% when the electron is in the K-shell, but that it will be found outside of that same sphere with a probability of 92% when it is in an L-shell.

THE HYDROGEN MOLECULAR ION

The hydrogen molecular ion comes into existence when two protons are held together by an electron. The latter's electrostatic attraction of the protons acts as a counterbalancing force to the protons' mutual repulsion. The stability of the molecular ion is due to the equilibrium established between these attractive and repulsive forces.

The experiment reveals that the average distance separating the protons in the ground state of the molecular ion is 1.06 Å; the dissociation energy of the ion amounts to 2.64 eV (electron volts). Strictly speaking, the wave mechanical study of this molecule is of the nature of a three-body problem since three particles are involved. By virtue of the fact that the protons have a mass far exceeding that of the electron, one may use the so-called Born-Oppenheimer approximation which simplifies many problems. It can be shown, actually, that it is possible to calculate the molecular wave function to a good approximation in many instances, by first studying the movements of the electrons in the field of the assumedly fixed nuclei, and only afterwards accounting for the movements of the nuclei. Such a procedure will be applied here. In the present case it amounts to a replacement of the molecular ion by a simplified model consisting of two fixed charges $+e$, denoted A and B and separated by a distance r_{AB}, and an electron trapped in the field generated by these two charges (Figure 5). The potential energy felt by an electron in point M will be

$$F(M) = -(e^2/r_{AM}) - (e^2/r_{BM}).$$

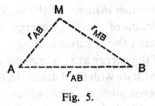

Fig. 5.

The wave equation for a stationary state of this model must be written then

$$- (h^2/8\pi^2 m)\, \nabla \Psi + (-(e^2/r_{AM}) - (e^2/r_{BM}))\, \Psi = E\Psi.$$

This equation was written out and solved by the Danish scientist Burrau in 1927. The structure of this equation depends on the distance between the two nuclei; the electron's energy values depend on this distance. Burrau recognized that, just like in the case of the hydrogen atom, here too the energy of the electron is quantized. In order to obtain the energy E of the molecule with the fixed nuclei, it suffices to add the electrostatic repulsion between the nuclei to the electronic energy. One finds

$$E_T = E + e^2/r_{AB}.$$

Burrau calculated the lowest possible value of this energy as a function of r_{AB}. The graph in Figure 6 represents the results of his calculation. The energy in this model passes through a minimum for a certain value r_e of the distance r_{AB}, such that rather curiously

$$r_e = 1.06 \text{ Å}.$$

According to the experiment, this value coincides precisely with the average distance separating the nuclei in the ground state of H_2^+. This was quite an encouraging result, showing on one hand the practical value of wave mechanics for the study of the chemical bond, and on the other hand that the nuclei are most probably separated by a distance close to r_e. It has become common

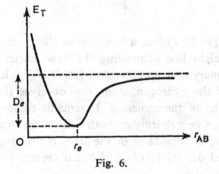

Fig. 6.

practice to call r_e the *equilibrium distance* of the nuclei, since in this model it corresponds with the minimum of electronic energy. When r_{AB} goes to infinity E_T asymptotically approaches the electronic energy of a hydrogen atom. The difference between E_T and the energy at r_e represents the energy to be supplied to the molecule in its ground state with fixed nuclei, in order to have it dissociate. This amount of energy has been given the name *electronic dissociation energy* and is denoted by D_e. Burrau's calculations gave $D_e = 2.7773$ eV, different by some 0.14 eV from the experimentally found total dissociation energy. It is easy to understand the cause underlying this difference. So far we considered the nuclei to be fixed in one position, although this is not actually the case. A more profound analysis of the problem by the methods of wave mechanics reveals, in fact, that even in the state of minimum energy, i.e. in the *ground state*, there exists a certain probability density of the distance r_{AB} separating the nuclei. This density reaches a maximum for $r_{AB} = 1.06$ Å. It decreases rapidly when r_{AB} departs from this value. It results in a certain movement of the nuclei with respect to each other, the vibrational motion, which remains present even at the absolute zero of the temperature scale, can never be stopped and has an energy of 0.14 eV. There is, then, a perfect agreement between theory and experiment. This agreement provides confidence in the solution of the previous equation and so in the wave function for the lowest eigenvalue, i.e. for the ground state of our model. Burrau also studied the form of the function $\Psi(M, t)$ in the special case that $r_{AB} = r_e = 1.06$ Å. The graph and the mapping in Figure 7 summarize his results. The graph represents the variation of the

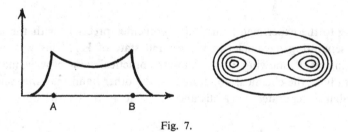

Fig. 7.

probability density $|\Psi(M, t)|^2$ as a function of the position of M, the latter situated on the straight line connecting the fixed nuclei A and B. We are dealing with a stationary state and therefore $|\Psi(M, t)|^2$ is independent of time. Like in the case of the hydrogen atom, one observes that this density has maxima at the nuclei of the molecule. It remains rather large between the nuclei, and then drops very sharply on both sides of the interval defined by the nuclei. The mapping is composed of the lines connecting the points where the above mentioned density $|\Psi(M, t)|^2$ has a certain constant value; such lines are called *electronic iso-density lines*. One may observe that the electronic cloud, i.e. the region where for all practical purposes the electron can be found, takes the shape of a cigar with the line connecting the nuclei serving as its axis and having its extreme values close to these nuclei. Such is the wave mechanical picture of the simplest of chemical bonds: the one-electron bond in the molecular ion H_2^+.

THE NATURE OF THE ENERGY OF A ONE-ELECTRON SYSTEM

Let us return to the wave equation of a stationary state

$$- (h^2/8\pi^2 m) \nabla \Psi + F\Psi = E\Psi.$$

The two members of this equation are now going to be multiplied from the left by the complex conjugate Ψ^* of the wave function, and the result thus obtained integrated over all space. We find then

$$\int \Psi^* (- (h^2/8\pi^2 m) \nabla) \Psi \, dv + \int \Psi^* F\Psi \, dv = E \int \Psi^* \Psi \, dv.$$

We have seen before, though, that Ψ is normalized to unity, so that

$$\int \Psi^* \Psi \, dv = \int |\Psi|^2 \, dv = 1.$$

This leads to

$$E = \int \Psi^* (- (h^2/8\pi^2 m) \nabla) \Psi \, dv + \int \Psi^* F\Psi \, dv. \tag{1.21}$$

The second term on the right of this Equation (1.21) may be written

$$\int \Psi^* F\Psi \, dv = \int F(M) |\Psi|^2 \, dv = \int F(M) \, dp, \tag{1.22}$$

where $|\Psi|^2 \, dv = dp$.

This integral is obtained by multiplying the potential energy of the electron in volume element dv around M, by the probability of its presence there, and subsequently summing these products after their multiplication by the corresponding volumes over all the elements constituting the space. In the sense of probability theory the integral represents the mathematical expectation of the potential energy, or rather, the *average* value, $\bar{F}$, of that energy. We have now

$$E = \int \Psi^* (- (h^2/8\pi^2 m) \nabla) \Psi \, dv + \bar{F}.$$

Looking again for analogies with classical mechanics, one is inclined to consider the first term as representing the average kinetic energy T. The equation becomes

$$E = \bar{T} + \bar{F}. \tag{1.23}$$

In order to keep our language consistent it suffices to associate the operator $-(h^2/8\pi^2 m) \nabla$ with the kinetic energy. This line of thought is justified by the fact that for a Coulomb potential acting on the electron

$$\bar{T} = -\tfrac{1}{2}\bar{F}.$$

This so-called *virial theorem* holds also true in wave mechanics. The result is that

$$E = \tfrac{1}{2}\bar{F}. \tag{1.24}$$

This very important equation shows that the total energy of a one-electron system equals one-half of the average potential energy. It clearly bears out the nature of the total energy. A system will be the more stable the more frequently its electrons stay in regions with a low potential. The origin of the bond in H_2^+ can be easily understood now, noticing that a proton H^+ approaching an atom H gives rise to a new region with a low potential in the vicinity of the former and between the nuclei, a region where the electron will often go to. It is the very existence of a region with a low potential between the nuclei which explains the stability of the molecular ion H_2^+.

4. Two-Electron Systems; the Helium Atom and the Hydrogen Molecule

THE WAVE MECHANICS OF THE SYSTEM

The problems posed by the one-electron systems H and H_2^+ could be dealt with successfully by replacing them with models with fixed nuclei. When several electrons are involved we have to rely on the wave mechanics of systems of particles, the principles of which we have yet to explain. To this end we take the example of a system with two electrons; the generalization to n electrons is then easy, afterwards. For simplicity's sake we will neglect the spin initially.

Let M_1 and M_2 be two points in space at time t. The presence of two electrons in space gives rise to a probability wave

$$\Psi(M_1, M_2, t)$$

such that the probability $dp_{1,2}$ of finding at time t electron number 1 in a volume element dv_1 around M_1, and simultaneously electron number 2 in a volume element dv_2 around M_2, shall be given by the relation

$$dp_{1,2} = |\Psi(M_1, M_2, t)|^2 \, dv_1 \, dv_2.$$

By definition, this wave is the solution of the equation

$$- (h^2/8\pi^2 m)\, \nabla_1 \Psi - (h^2/8\pi^2 m)\, \nabla_2 \Psi + F(M_1, M_2, t)\, \Psi =$$
$$= (h/2\pi i)\, \partial\Psi/\partial t,$$

where ∇_1 and ∇_2 are the Laplace operators differentiating Ψ at respectively M_1 and M_2, and $F(M_1, M_2, t)$ the classical potential energy acting on the electrons at time t when one is at M_1 and the other at M_2. The principles of quantization and spectral decomposition are applicable. In the case of a stationary state the wave takes the form

$$\Psi = a(M_1, M_2)\exp(2\pi i v t)$$

with $E = hv$.

E is the total energy of the electrons; the equation becomes

$$- (h^2/8\pi^2 m)\, \nabla_1 \Psi - (h^2/8\pi^2 m)\, \nabla_2 \Psi + F(M_1, M_2)\, \Psi = E\Psi.$$

One should clearly understand that the wave is associated with the ensemble of electrons; one cannot associate an individual wave with an individual electron in the system. Likewise, the energy is only distinctly defined for the ensemble of electrons. It is impossible to speak of the energy of either electron.

The wave mechanics of systems of electrons postulates a fourth principle: electrons are indistinguishable. Let us specify how that should be understood. We had already

$$\Psi(\overset{1}{M_1}, \overset{2}{M_2}, t),$$

the wave associated with the simultaneous search for electron 1 in M_1 and 2 in M_2. The function

$$\Psi(\overset{1}{M_2}, \overset{2}{M_1}, t)$$

shall be the wave associated with the simultaneous presence of 1 in M_2 and 2 in M_1. In the argument of the function this notation classifies the finding places of the electrons in the numerical order of those particles. The probability $dp_{1,2}$ of simultaneously finding 1 in dv_1 around M_1, and 2 in dv_2 around M_2, is written like done before

$$dp_{1,2} = |\Psi(M_1, M_2, t)|^2\, dv_1\, dv_2,$$

and similarly the probability finding 1 in dv_2 and 2 in dv_1 is

$$dp_{2,1} = |\Psi(M_2, M_1, t)|^2\, dv_2\, dv_1.$$

The principle of *indistinguishability* imposes the necessity of $dp_{1,2} = dp_{2,1}$. This is because a physicist without tools to distinguish between individual electrons, experiences the two situations as equivalent. His inescapable conclusion is that

$$|\Psi(M_1, M_2, t)|^2 = |\Psi(M_2, M_1, t)|^2,$$

from which follows, after introduction of the simplifying assumption that one deals with real wave functions only, that

$$\Psi(M_1, M_2, t) = \pm \Psi(M_2, M_1, t).$$

The wave is *symmetrical* or *anti-symmetrical* with respect to a *permutation* of the two points. It should be pointed out that when

$$\Psi(M_1, M_2, t) = - \Psi(M_2, M_1, t),$$

and substitution of

$$M_1 = M_2 = M,$$

this equality becomes

$$\Psi(M, M, t) = - \Psi(M, M, t),$$

and so

$$\Psi(M, M, t) = 0.$$

The probability of finding the two electrons in the same volume element in space equals zero for a state described by an anti-symmetrical wave.

THE HELIUM ATOM

It is now possible to discuss the helium atom according to these principles when the nucleus is replaced by a charge $+2e$, fixed in space (Figure 8). It is straightforward to understand that in this case

$$F(M_1, M_2) = - (2e^2/r_{M1}) - (2e^2/r_{M2}) + (e^2/r_{M1M2}).$$

The wave function assumes the form

$$- (h^2/8\pi^2 m) \nabla_1 \Psi - (h^2/8\pi^2 m) \nabla_2 \Psi + $$
$$+ (- (2e^2/r_{M_1}) - (2e^2/r_{M_2}) + (e^2/r_{M_1M_2})) \Psi.$$

This equation has been solved nearly rigorously, by E.A. Hylleraas in 1930 [7],

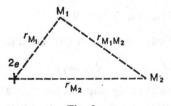

Fig. 8.

but the theorem of the existence of any solution was not solved until 1951 [8]. For the lowest energy of the helium atom Hylleraas found -78.6 eV, in perfect agreement with the experimental value. The function $\Psi(M_1, M_2, t)$ of the ground state is symmetrical. The next higher eigenvalue corresponds with what has become known as the first excited state of the atom. It has been found that this second eigenvalue belongs to an anti-symmetrical function.

FIRST APPEARANCE OF THE CONCEPT OF THE LOGE

Two pictures of the helium atom will now be compared: the one given by wave mechanics and the one provided by Bohr's theory. In the latter the ground-state of the helium atom shows two electrons circling in one orbit, denoted by K; the first excited state must then have one electron in this K-orbit and the other in a L-orbit. Electrons being indistinguishable in wave mechanics, the two must, on the average, play exactly the same role in any given state of the atom. One cannot make the distinction between K-, L-, ... electrons. When, for instance, the average distance $\bar{r} = \int r\, \mathrm{d}p$ is calculated for the excited state on the basis of the symmetry properties of the wave Ψ, one finds the same value for the two electrons. The modern picture of the atoms seems to be a good deal different from the more ancient picture. However, certain analogies can still indicated. We will place the nucleus in the center of a sphere with an arbitrary radius R, and with the aid of the wave function belonging to the first excited state we will calculate the probability P of finding one and only one electron within this sphere. When R is very small, P will be very small, simply because there is little chance of finding an electron in so small a sphere. When R tends to infinity the sphere occupies almost the entire space and is bound to contain both electrons. The probability of finding a single electron tends to zero in this case. This means that P should possess at least one maximum. The diagram in Figure 9 shows the result of such a calculation [9a–d]. One sees that for $R = 1.7a$ the probability of finding one and only one electron within the sphere,

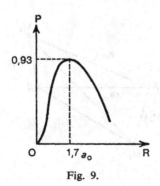

Fig. 9.

equals 93%. Naturally, at a moment that one single electron is within the sphere, the other cannot be but outside of it. In a helium atom in its first excited state, one electron will frequently be found within the sphere while the other is outside of it. One could say now that the domain within the sphere with a radius of 1.7 a constitutes loge K, and the remainder of space loge L. The conclusion may be drawn that there exists a high degree of probability (93%) that one electron will be found in loge K and the other in loge L, when the helium atom is in its first excited state. It is evident that this formulation respects the indistinguishability of the electrons. There is no way to speak about K- or L-electrons. Such labels will from now on refer only to distinct domains of space.

THE HYDROGEN MOLECULE

We will stay within the limitations of the Born-Oppenheimer approximation. The treatment of the hydrogen molecule is then reduced to a two-body problem: two electrons in the field of the two fixed nuclei A and B (Figure 10). The wave equation is

$$- (h^2/8\pi^2 m) \nabla_1 \Psi - (h^2/8\pi^2 m) \nabla_2 \Psi +$$
$$+ (e^2/r_{M_1 M_2} - e^2/r_{1 MA} - e^2/r_{M_1 B} - e^2/r_{M_2 A} - e^2 r_{M_2 B}) \Psi = E\Psi.$$

This equation has been solved almost rigourously by James and Coolidge in 1933 [10]. The graph in Figure 11 shows how the two lower eigenvalues E_{T1} and E_{T2} of the total energy, vary with r_{AB}: $E_T = E + e^2/r_{AB}$. In this case the graph of the energy of the ground state presents exactly the same kind of picture as in the case of the molecular ion H_2^+. The minimum of the energy appears at $r_e = 0.74$ Å. Again, this distance coincides with the experimentally determined average distance of the nuclei. The value calculated for D_e is 4.70 eV; experiment gives 4.72 eV. Once again the conditions seem favourable for a linking of applied rationalism and technical materialism. Recent calculations and measurements have given even more substance to the similarity [11]. The

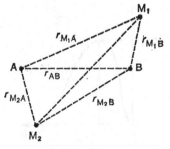

Fig. 10.

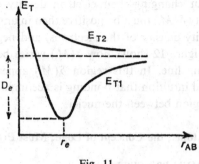

Fig. 11.

following results emerged:

$$\text{for } r_e \begin{cases} \text{experimental value } 0.74116 \text{ Å} \\ \text{theoretical value } 0.74127 \text{ Å} \end{cases}$$

$$\text{for } D_e \begin{cases} \text{experimental value } 4.7466 \pm 0.0007 \text{ eV} \\ \text{theoretical value } 4.7467 \text{ eV}. \end{cases}$$

The graph of the variation of the probability density of an electron, or the electron density, along the line connecting the nuclei resembles that shown in Figure 7 for H_2^+. The mapping of the iso-density lines is also of the same type. Just as in the case of H_2^+, the electron cloud of H_2 is cigar shaped, but since two electrons, simultaneously exerting their Coulomb attraction, can more effectively challenge the repulsive force between the nuclei than one single electron, the nuclear distance in H_2 is shorter than in H_2^+; the cigar is shorter and consequently thicker. The nature of the bond is always the same: the electrons tend to bring the nuclei together but the latter repel each other until a certain overall equilibrium has been reached. The graph corresponding with the second eigenvalue, i.e. with an excited electronic level, exhibits a rather different behaviour. As the energy decreases monotonously with increasing separation of the nuclei, it defines an unstable state of the hydrogen molecule. One also finds, like in the case of helium, that the ground state wave function is symmetrical, whereas the excited state is associated with an anti-symmetrical function. We know that the probability to find two electrons simultaneously in the same domain in space is very small for an anti-symmetrical wave function. The electrons find it difficult to stay closely together between two nuclei; on a qualitative level one can thus understand why such a state will be unstable. Let us return now to the ground state. It is possible to define for any molecule the *differential density* $\delta(M)$ [12] in every point M. This differential density is the difference between the true electronic density $\varrho(M)$ at point M, and the density $\varrho^f(M)$ which would prevail in the same point, had the atoms been superimposed without changing their electron density during the approach. The differential density $\delta(M)$ must be positive then in any point where bonding adds to the probability density of the electrons, and negative in the opposite case. The graph in Figure 12 shows how $\delta(M)$ varies between the nuclei and along their connecting line. In this region $\delta(M)$ appears to be positive. It confirms our chemical intuition that bonding is accompanied by the migration of electrons to the region between the nuclei.

5. Many-Electron Atoms; the Concept of Loge; Most Probable Configuration

ELECTRON SPIN AND PAULI PRINCIPLE

A great many experimental facts not suitable for a further analysis at the present stage, led physicists to believe that the electron possesses an *intrinsic kinetic momentum*. The designation *kinetic momentum* usually refers to the momentum of a moving body with respect to a fixed point. Intrinsic kinetic momentum should be understood to be the electron's kinetic momentum with respect to itself. In order to explain the existence of such a momentum one should imagine the electron to be rotating about itself, just like the Earth rotates about one of its axes. It is difficult to present an equivalent picture in terms of wave mechanics. We were not previously confronted with this delicate problem. We will assume that an electron possesses an intrinsic momentum to be called *spin*, having the length $S(S+1)^{1/2} h/2\pi$; $S=\frac{1}{2}$. We shall further assume that the length of the projection of this vector on any axis shall be either $+\frac{1}{2}h/2\pi$ or $-\frac{1}{2}h/2\pi$. These postulates allow of a correct interpretation of all the experimental facts associated with the spin. In non-relativistic wave mechanics the spin is often introduced by taking into consideration that the wave Ψ no longer solely depends on the points where one locates the electrons, but also on the lengths of the spin projections with which one wishes to find them. In the case of a single electron the wave

$$\Psi(M, + \tfrac{1}{2}h/2\pi, t)$$

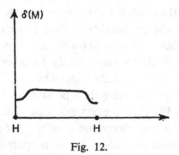

Fig. 12.

is assigned to it when the spin projection is positive, and the function

$$\Psi(M, -\tfrac{1}{2}h/2\pi, t)$$

with a negative spin projection. Putting it differently, $|\Psi(M, +\tfrac{1}{2}h/2\pi, t)|^2$ represents the electron density for positive projection, at point M, and similarly $|\Psi(M, -\tfrac{1}{2}h/2, t)|^2$ idem but for a negative spin projection. As a straightforward generalization,

$$\Psi(\underbrace{M_1, +\tfrac{1}{2}h/2\pi}_{1}, \underbrace{M_2, -\tfrac{1}{2}h/2\pi}_{2}, t)|^2$$

represents the simultaneous density of electron 1 at M_1 with positive spin projection, and electron 2 at M_2 with negative spin projection. The wave

$$\Psi(M_1, \sigma_1, M_2, \sigma_2, ..., M_n\sigma_n, ..., t)$$

should be associated then with a population of electrons which one finds in the points $M_1, M_2, M_3, ..., M_n$ respectively, with the respective spin projections $\sigma_1, \sigma_2, \sigma_3, ..., \sigma_n$.

Returning to the two-electron case,

$$|\Psi(\underbrace{M_1, \sigma_1,}_{1} \underbrace{M_2, \sigma_2}_{2}, t)|^2$$

stands for the simultaneous density of electron 1 in M_1 with projection σ_1 and electron 2 in M_2 with projection σ_2. In the same way, the notation

$$|\Psi(\underbrace{M_2, \sigma_2,}_{1} \underbrace{M_1, \sigma_1,}_{2} t)|^2$$

stands for the simultaneous density of electron 2 in M_1 with projection σ_1, and electron 1 in M_2 with projection σ_2. According to the principle of indistinguishability, the equality

$$|\Psi(\underbrace{M_1, \sigma_1,}_{1} \underbrace{M_2, \sigma_2,}_{2} t)|^2 = |\Psi(\underbrace{M_2, \sigma_2,}_{1} \underbrace{M_1, \sigma_1,}_{2} t)|^2$$

must hold, and so

$$\Psi(M_1, \sigma_1, M_2, \sigma_2, t) = \pm \Psi(M_2, \sigma_2, M_1, \sigma_1, t).$$

Pauli showed that in order to give a correct account of the various experimental data, the $+$ in this two-valued equality has to be excluded in the case of an ensemble of electrons. This so-called *Pauli principle* generally amounts to the assertion that a wave function associated with a population of electrons, is anti-symmetrical with respect to a permutation of the space coordinates (M_1 and M_2) and the spin coordinates (σ_1 and σ_2) belonging to the two electrons. This principle has direct physical consequences. Since

$$\Psi(M_1, \sigma_1, M_2, \sigma_2, t) = -\Psi(M_2, \sigma_2, M_1, \sigma_1, t)$$

and substituting $M_1 = M_2 = M$; $\sigma_1 = \sigma_2 = \sigma$, one finds that $\Psi(M, \sigma, M, \sigma, t) = -\Psi(M, \sigma, M, \sigma, t)$ which can be true only when $\Psi(M, \sigma, M, \sigma, t) = 0$.

The probability of finding two electrons with the same spin projection in the same domain of space, equals zero.

It can be shown that in the two-electron case the wave function may be written as the product of a function of the space coordinates, $\phi(M_1, M_2, t)$, and a function of the spin coordinates, $\sigma(\sigma_1, \sigma_2)$:

$$\Psi(M_1, \sigma_1, M_2, \sigma_2, t) = \phi(M_1, M_2, t)\,\sigma(\sigma_1, \sigma_2).$$

It was found that two types of functions $\phi(M_1, M_2, t)$ exist: the symmetrical functions $\phi^S(M_1, M_2, t)$ and the anti-symmetrical functions $\phi^A(M_1, M_2, t)$. In order to comply with the Pauli principle, a space function has to be multiplied by a spin function of the opposite summetry. In other words, the following types of products have to be formed:

$$\phi^S(M_1, M_2, t)\,\sigma^A(\sigma_1, \sigma_2) \quad \text{and} \quad \phi^A(M_1, M_2, t)\,\sigma^S(\sigma_1, \sigma_2).$$

In addition, it could be shown that there exists only one anti-symmetrical function σ^A, but as many as three functions σ^S. A state represented by σ^S is therefore called a *singulet*, a state represented by σ^A a *triplet* by virtue of the fact that the spin function introduces a three-fold degeneracy into the total wave function. For all states the probability density of two electrons with equal spin projections equals zero because of the Pauli principle. For triplet states, though, this density equals zero even for electrons with opposite spin projections, due to the anti-symmetry of the function ϕ^A, as was shown earlier on. It is sometimes said that two electrons with the same spin projections are coupled, and it is even suggested that such a coupling reflects an attraction. The reader should be aware of the fallacious nature of such usage. The existence of electrostatic forces makes that electrons always repel one another. There are certain cases in which the probability of finding two electrons with opposite spin projections in the same small volume, does not equal zero. However, when two electrons have parallel spins, the probability is zero without exception.

Two electrons with opposite spin projections allow themselves to be neighbours, in certain cases. Two electrons with parallel spin projections never cohabitate. Never is it a matter of love, only of a lesser hatred.

THE CONCEPT OF LOGE

The problems that arose in the case of the helium atom are also present when we study other atoms. Specifically, one finds that, owing to the symmetry properties of the wave functions, the average distance of an electron from the nucleus is the same for all electrons. It remains as incorrect as in the case for helium, to distinguish between K-, L-, M-, ... electrons. Through a generalization of the concept of *loge*, a geometrical description bearing some resemblance to the picture of the old quantum theory can be made. Manipulating wave functions of heavy atoms in their ground states, one finds that it is always possible to indicate a sphere centered in the nucleus, within which the probability to find two and only two electrons with opposite spins is high; that is the K-loge. Around this K-loge it is possible to find a spherical corona in which there exists a high probability to find 4 electrons with spin $+\frac{1}{2}h/2\pi$ and 4 others with spin $-\frac{1}{2}h/2\pi$: that is the L-loge. The M-, N-, O-,... loges can be defined in a similar manner. Let us examine a particular ion, F^-. The space occupied by this atom may be divided into a K-loge with a radius of 0.35 a where the probability to find two electrons with opposite spins is 0.8, and a L-loge (the remainder of space) in which a high probability prevails to find 4 electrons with spin $+\frac{1}{2}h/2\pi$ and 4 with spin $-\frac{1}{2}h/2\pi$. Table I gives the radii of some loges.

We will now divide the volume of a loge by the highly probable number of electrons contained in it. This results in a certain volume v that gives us an impression of the average room occupied by an electron when it resides in that particular loge. Then we could calculate the average electrostatic potential p experienced by an electron in that loge from the nucleus and the other electrons. It turns out that between p and v the following relation holds [14]

$$p^{3/2}v = \text{Constant},$$

the Constant being the same for all loges and all atoms. There is thus a kind of Boyle-Mariotte relationship between the volume occupied by the electron, and the electrostatic 'pressure' acting on it.

THE MOST PROBABLE ELECTRON CONFIGURATION IN AN ATOM

Let us examine the electronic system represented by the stationary wave

$$\Psi(M_1, \phi_1, M_2, \sigma_2, ..., M_n, \sigma_n).$$

TABLE I [13]

Radii of spheres on which loges terminate (atomic units)
(Atomic or ionic symbols)

Type of loge	Be	F^-	Al^{+2}	Ca^{+2}	Rb^+
K	1.12	0.35	0.22	0.13	0.06
L				0.64	0.26
M					1.05

The collection (or collections) of points M_1, M_2,..., M_n corresponding with the highest maximum of the modulus of this wave bears the name *most probable configuration* of electrons in the system. In these points the simultaneous probability density reaches its greatest value. In Figure 13 it is shown [15] how this density presents itself in respectively beryllium, boron and carbon in their states of minimum energy with one electron having spin $-\frac{1}{2}h/2\pi$ plus, respectively, three four and five electrons having spin $+\frac{1}{2}h/2\pi$. Solid circles indicate places where electrons with positive spin projections may be expected, open circles the same for electrons with negative spin projections. One should notice that always in this kind of configuration, two electrons with opposite spins are found in the vicinity of the nucleus. The K-loge, symbolized by a circle in Figure 13, always contains two and only two of such electrons once the most probable configuration has been established. The L-loge (the remainder of space) contains, depending on the particular case, two, three or four electrons with the same spin projection and forming the largest possible angles between themselves and the nucleus. Once again, this is a consequence of the Pauli principle. Thanks to the concepts of loge and most probable configuration, we eventually succeeded in associating the arrangement of the electrons in atoms with geometrical representations compatible with the notions of wave mechanics.

6. The Concept of the Chemical Bond

SHORT HISTORY; DEFINITION OF VICINITY OF A NUCLEUS

John Dalton (16) was the first to clearly propose the idea that each chemical element should correspond with one particular kind of atom, and that a well-defined chemical compound should be the result of the union between an atom of one element and one, two, three,... atoms of another element. This thought proved very fruitful indeed, and before long the importance of the arrangement of the atoms in a molecule was surmized. The discovery of isomers, i.e. chemically distinct molecules containing exactly the same numbers and kinds of atoms, demonstrated that the arrangement of the atoms plays a decisive role in determining the properties of a molecule. Therefore, it has become essential to distinguish between *bonded* and *non-bonded* atoms. In 1916 G. N. Lewis laid the foundations of an electronic theory of the chemical bond [17] by

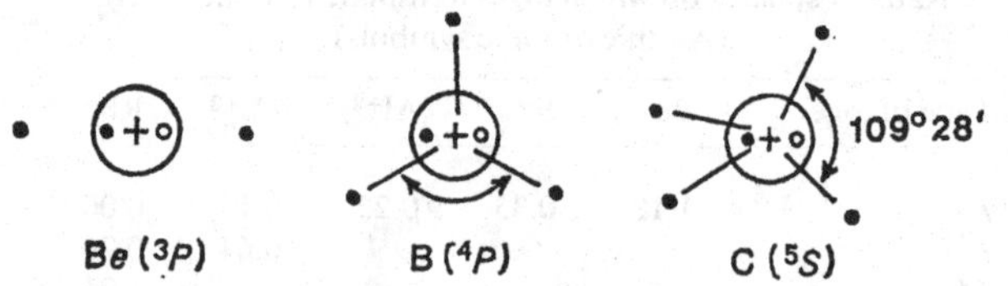

Be (^3P)   B (^4P)   C (^5S)   109° 28'

Fig. 13.

assuming that a single bond is the result of a sharing by two atoms of two electrons engaging in the formation of a chemical bond. It became possible and necessary to make a distinction between various modes of bonding: *ionic* bond, *covalent* bond, *coordinative* bond (or *donor-*, or *semi polar* bond) [18a–d]. With the advent of wave mechanics, his concepts faced formidable difficulties. In fact, we assume nowadays that a molecule is constituted by an ensemble of nuclei, but we have remained unable to trace the electrons' trajectories. The nuclei exert mutually repulsive forces. They are only prevented from dispersion by the attractive forces brought to bear by the electrons. Under such circumstances it is difficult to distinguish between bonded and non-bonded atoms: any two nuclei are always interacting with the entire electronic cloud. It is not permitted to associate two particular electrons with a bond (after all hard to define), since the principle of indistinguishability leads to an anti-symmetric wave function and imposes on all electrons the duty to play, on the average, the same role.

In order to solve this difficult problem we shall first try to analyze the nuclear topology of molecules. Fortunately, the distances between nuclei in molecules can be measured. Spectroscopic methods have a very high precision for simple molecules; the error is of the order of 0.001 Å. For more complex molecules X-ray diffraction should have the preference. This less accurate method leaves an uncertainty of 0.01–0.05 Å in normal cases. A surprisingly high precision can be obtained with highly symmetric systems of atoms: for instance, in diamond the shortest $C-C$ distance is calculated at $1.54452 \pm \pm 0.00014$ Å. Let us now disregard chemical formulae and take a series of molecules containing carbon atoms with known $C-C$ distances, i.e. the distances between any two carbon atoms in the molecule. We will take the statistics of the number of $C-C$-bonds with a length within a given interval [19]. Figure 14 shows the outcome of such a count. One observes the appearance of a first zone comprising the distances between 1.2 and 1.6 Å; then follows a 'forbidden' zone from 1.6 to 2.2 Å beyond which any distance may be encountered. The existence of the 'forbidden' interval makes it possible to define vicinity. We shall call two carbon atoms vicinal when they are separated by a distance lying within the first 'allowed' zone, i.e. between 1.2 and 1.6 Å. This method may be generalized for other atoms, and by doing so the basis can be laid for a nuclear topology by defining the ensemble of vicinal nuclei of one particular nucleus in a molecule. One could say, should one wish so, that vicinal nuclei are bonded together. Obviously, in this way the chains, indicated by the bond-dashes in classical formulae, reappear almost exactly with this method.

CORE LOGES AND BOND LOGES

A second problem still faces us: how to find some way of assigning electrons to

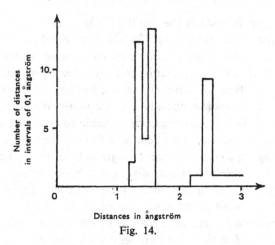

Fig. 14.

nuclei? The concept of loge allowed us to replace the classical shells by domains of space, and our intuition tells us that the same concept may prove useful again. Take a look at the lithium molecule Li_2 which contains six electrons. Around each nucleus a sphere with radius R is constructed, and the search is for the value of R optimizing the probability to find two and only two electrons with opposite spins within each one of these spheres (Figure 15). This maximum appears at $R=1.53$ a with a value of 90% [20]. This radius is evidently the same as the one of the K-loge in the lithium atom. In the Li_2 molecule the K-loges of the two lithium atoms appear nearly unaltered. Those two spheres with a radius of 1.53 a each will now be said to constitute the *core loges* of the lithium molecule. In the remaining molecular space between these loges, the probability to find two and only two electrons with opposite spins equals 97% [21]. This remaining molecular space constitutes a two-electron bond loge. In this way the concept of loge permits us to replace the two electrons assigned by Lewis to a single covalent bond by a domain of space where a high probability prevails of finding two and only two electrons with opposite spins. Once again, we label domains of space instead of electrons, since the latter are indistinguishable. It could be objected that the concept of loge is somewhat spurious, since it is not very clear what else induced us to the trisection of the

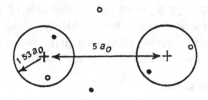

Fig. 15.

Li_2 molecule into three loges than the desire to arrive at something similar to Lewis' presentation. Fortunately, it is possible to remove the artificiality by referring to certain aspects of information theory. In fact, it is possible to define by application of this theory, the best partition into loges as that one that confers upon the molecule a maximum of information; i.e. the one minimizing the missing information or the *a priori* indetermination [22]. The concept of loge, in the words of K. Jørgensen [22], introduced for the first time something 'aristic' into wave mechanics. (The neologism aristic is derived from the Greek ἄριστος, the better one.)

THE DIFFERENTIAL DENSITY IN SOME DIATOMIC MOLECULES

We used the differential density function $\delta(M)$ before, in an attempt to understand some of the characteristic features of the single bond par excellence: that in H_2. We were dealing then with a homonuclear molecule, i.e. a molecule containing nuclei of one kind only. Figure 16 gives the variation of that function $\delta(M)$ along the line connecting the nuclei of a heteronuclear molecule, LiH or lithium hydride [23a, b]. The function is positive near the hydrogen nucleus and becomes negative near the lithium nucleus. The formation of the molecule out of the atoms is accompanied by a simultaneous transfer of electrons by which the region around the hydrogen is enriched at the expense of the region around the lithium. This result clarifies the origin of the polarity of this molecule, sometimes indicated by the symbolism $Li^{\delta+}-H^{\delta-}$. Figure 17 gives a schematic account of the results from a calculation of this same function $\delta(M)$ in one of the planes containing the nuclei of an oxygen molecule [24], where chemists usually assume the presence of a double bond, or, translated into the theory of loges, where they suggest the existence of a domain between the cores of the oxygens with a prevailing high probability to find two and only

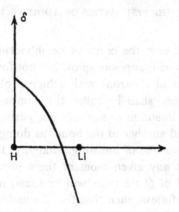

Fig. 16.

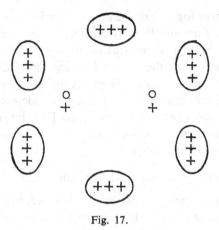

Fig. 17.

two pairs of electrons with opposite spins. The illustration in Figure 17 remains restricted to the domain where $\delta(M)$ is positive. Between the domains $\delta(M)$ is negative. One observes a negative differential density along the nuclear axis, and on both sides of this line regions enriched with electrons. The origin of this phenomenon can be easily understood by remembering that two electrons with the same spin cannot stay together in the same domain in space. Of the two pairs of electrons two will necessarily have a spin $+\frac{1}{2}h/2\pi$ and the other two $-\frac{1}{2}h/2\pi$; this explains the tendency of the electrons to concentrate on the sides of the nuclear axis: there is 'too little room' for the electrons with like spins between the cores. Finally, it should be noted that the existence of enriched domains close to the plane of the oxygens but outside the connecting line of the nuclei, is not unlike the bent valence bonds used by chemists to indicate double bonds.

SATURATED MOLECULES; THE SYSTEMATICS OF ADDITIVE MOLECULAR PROPERTIES; ISOMERIZATION ENERGY

We are aware that between the cores of neighbouring light atoms there is room for two electrons with opposite spins, but not for more than two. From now on N, the number of electrons with a high probability of being found outside of the core loges, shall be called the number of bonding electrons. However, it is not our intention to say that one particular electron should be assigned to the core and another to the bond, as doing so would amount to a contradiction of the principle of indistinguishability. It should be interpreted so as to mean that at any given moment there prevails a high probability that $Q-N$ of the total of Q electrons will be found inside the core, and the rest on the outside. It follows then that for N equalling twice the number P of neighbouring pairs of cores, one has to expect that between every pair of neighbouring cores a loge with two electrons will be found. Such a loge may be said to define a *localized* (i.e. between two cores) *two-electron bond*. This is how to explain the concept of single bond in quantum language.

Molecules with $N=2P$ form the group of the so-called saturated molecules. The *saturated hydrocarbons* belong to this group of molecules. Figure 18 shows

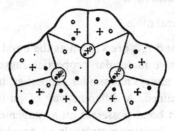

Fig. 18.

the partitioning into loges for the propane molecule. From the early years of our century onwards, attention was given to the additive properties of matter. In 1910 already, P. Pascal [25] put forward excellent systematics of additive molecular properties permitting the calculation within a fair approximation, of the magnetic susceptibility of a molecule by summation of the empirical increments associated with its constituents. Several additive properties may be mentioned, such as specific heat, heat of formation, heat of combustion, refraction, magnetic rotation, electrical moment etc. Table II presents some specific values with which it is possible to calculate the susceptibilities and the magnetic rotations with an accuracy of 1%, of molecules built from the units mentioned [26].

TABLE II

Susceptibility $[\chi]_i$ and magnetic ro-
totation $[\varrho]_i$ of some bonds; $[\chi]_i$ in
electromagnetic cgs-units; $[\varrho]_i$ in
μ.r. at 5780 Å and 20 °C

Bonds	$[\chi]_i$	$[\varrho]_i$
C—C	− 2.9	18.5
C—H	− 4.25	27.2
C—F		20
C—Cl	−20	131
C—Br	−29.3	267
C—I	−46.1	566
C—N	− 3.4	47
C—O	− 4.1	14.2
N—H	− 5.7	20
O—H	− 5.6	28

In a similar way one can calculate the energy of formation of a saturated hydrocarbon, i.e. the difference between the summation of the energies of the free atoms and the energy of the molecule, using the formula

with
$$N_{CC} = E_{CC} + N_{CH}E_{CH}$$

and
$$E_{CC} = 98.75 \text{ kcal mole}^{-1}$$

$$E_{CH} = 83.6 \text{ kcal mole}^{-1}.$$

The specific values E_{CC} and E_{CH} are known as the bond energies. The existence of such systematic additive molecular properties suggests that a single bond between two given cores always has similar properties, irrespective of the molecule in which it is contained. A different way of putting this is that a loge associated with a certain bond possesses an arrangement of its electrons that varies only very slightly from one molecule to another. This idea is further confirmed by the fact that in saturated molecules the distance between two nuclei of a given kind is virtually constant (see Table III). It is easy to show [27a–b] that a quantity A represented by a one-electron operator, may be decomposed approximately into the sum of contributions A_i from all the loges:

$$A \approx \sum_i A_i.$$

TABLE III

	Bondlengths in Å	
	C—C	C—H
Methane		1.091
Ethane	1.543	1.102
Propane	1.54 ± 0.02	
Isobutane	1.54 ± 0.02	
Cyclohexane	1.54 ± 0.02	

When, on the other hand, the operator is of the two-electron type, as is the case with energies depending on the interaction between two electrons, the quantity A may be decomposed into the sum of the contributions from the loges plus the sum of the terms A_{ij} representing the interaction between loge i and loge j:

$$A = \sum_i A_i + \sum_{i<j} A_{ij}.$$

Thus there is a direct justification for the systematic additive molecular properties of one-electron quantities when one assumes an invariable arrangement of the electrons in the two-electron loges extending between the two given cores.

The situation is more complicated for two-electron quantities since these also require that every loge has a similar environment. Therefore, differences appear when environments are different. The *isomerization energy* is a result of that effect. Compare normal butane

$$
\begin{array}{c}
\text{H} \quad \text{H} \quad \text{H} \quad \text{H} \\
| \quad | \quad | \quad | \\
\text{H} - \text{C} - \text{C} - \text{C} - \text{C} - \text{H} \\
| \quad | \quad | \quad | \\
\text{H} \quad \text{H} \quad \text{H} \quad \text{H}
\end{array}
\qquad \text{with iso-butane} \qquad
\begin{array}{c}
\text{H} \quad \text{H} \quad \text{H} \\
| \quad | \quad | \\
\text{H} - \text{C} - \text{C} - \text{C} - \text{H} \\
\quad \text{H} \quad | \quad \text{H} \\
\quad \text{H} - \text{C} - \text{H} \\
\quad | \\
\quad \text{H}
\end{array}
$$

We neglect interactions between loges belonging to different cores, considering that Coulomb forces are inversely proportional to the distance. The zones between the dashed lines in the preceding formulae indicate loges with different environments. In the zone of *n*-butane one finds an interaction energy equal to

$$2A_{CC,CC} + 2A_{CH,CH} + 8A_{CC,CH}.$$

In the zone of iso-butane the corresponding energy is

$$3A_{CC,CC} + 3A_{CH,CH} + 6A_{CC,CH}.$$

The difference in energy of formation between these two molecules is therefore

$$\Delta E = A_{CC,CC} + A_{CH,CH} - 2A_{CC,CH},$$

which is the isomerization energy. Calculating the same difference for *n*-pentane and 2,2-dimethyl-propane, one finds

$$\Delta E' = 3A_{CC,CC} + 3A_{CH,CH} - 6A_{CC,CH},$$

and so

$$\Delta E' = 3\Delta E.$$

This later result is in very good agreement with the experimental values

$$\Delta E = 1.7 \pm 0.1 \ \text{kcal mole}^{-1}$$

and

$$\Delta E' = 4.7 \pm 0.1 \ \text{kcal mole}^{-1}.$$

COVALENT BOND; COORDINATIVE BOND

The distinction between covalent- and coordinative bond is brought about by consideration of the likely mechanism through which these bonds are formed. When we are interested in the bonding in FH we can conceive the formation of the bond to begin with a fluorine atom and a hydrogen atom; Lewis suggests the following notation:

$$\overset{\cdot\cdot}{\underset{\cdot\cdot}{:F}}\cdot\ +\ \cdot H \rightarrow\ \overset{\cdot\cdot}{\underset{\cdot\cdot}{:F}}\!:H.$$

It is tempting to say that the bond is brought into existence by the sharing of two electrons by two atoms, each atom contributing one electron. This mechanism defines the covalent bond.

When subsequently we focus our attention on the B—N-bond in borazane, H_3BNH_3, the following scheme is attractive:

$$H-B\overset{\displaystyle H}{\underset{\displaystyle H}{\Big\langle}}\ +\ :N\overset{\displaystyle H}{\underset{\displaystyle H}{\Big\langle}}-H\ \rightarrow\ H-B\!:\!N-H$$

Again, two electrons are jointly possessed, but in this case the two are supplied by the same atom. This atom, nitrogen in the present example, is called a donor whereas the other atom, borium, is called an acceptor. It can now be said that a *coordinative bond* has been formed, visualized by

$$H-B\ \leftarrow\ N-H$$

Other accepted names are donor-acceptor bond, semi-polar bond or dative bond. One discovers easily the problems posed by such a description. In the first place, there may be hesitation to choose between several mechanisms which for the same bond may lead to different definitions. In the second place, there is no assurance that the eventual arrangement of the electrons as suggested by a mechanism, will remain that way: there is always the possibility of a rearrangement. We can sense the importance of this latter phenomenon by examining the function of the ammonium ion in the reaction

$$H-N\!:\ +\ H^+\ \rightarrow\ \left[H-N\rightarrow H \right]^+$$

which invites us to consider one of the four bonds in this ion as a coordinative bond, although the tetrahedral structure and perfect symmetry of this entity forces us to assume an identical role of the electrons in the four bonds, and consequently their identical characters.

We have known for some time that it is not possible to assign particular electrons to a bond. The theory of loges offers a solution to this problem [28]. It will be given for the simplest cases only. The schemes in Figure 19 show the correct division into loges of borazane and methylamine. Chemists write it as follows

$$H_3B \leftarrow NH_3 \quad \text{and} \quad H_3C - NH_2 .$$

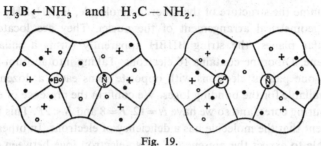

Fig. 19.

The most probable distribution shall be such a distribution of electrons in a molecule that in each loge precisely that number of electrons will be found which one could expect to find according to the highest probability in that loge. It is obvious that with such a distribution the loge BN in borazane lies between a group of loges BH_3 with zero charge (also considering the nuclei) and a group of loges NH_3 with charge $+2e$, whereas the loge CN in methylamine lies between groups of loges with charge $+e$. We will say that there is a *dative* bond in the former, and a *covalent* bond in the latter case. This definition obviously introduces no ambiguities since it depends on the final structure of the molecule only, and not on some mechanism of formation. It also respects the principle of indistinguishability. It should be noted that an examination of the charges present in the loges, allows of an *a priori* impression of the respective polarities of the bonds. In methylamine the CN loge is located between two groups of loges with equal charges $(+e)$ and between a core of carbon with a charge $+4$ and a core of nitrogen with a charge $+5$; attention should be given to the fact that the center of gravity of the electronic density in the CN loge is a little closer to the core of the nitrogen than to that of the carbon, so that the polarity of the bond may be represented by the symbol $C \cdot \leftrightarrow \cdot N$. In borazane the BN loge is located between a BH_3 group with zero charge and a NH_3 group with charge $+2e$, and between a core of B with charge $+3e$ and a core of nitrogen with charge $+5e$. There are two reasons, then, why the center of gravity is closer to the nitrogen. The polarity will be similar though more pronounced than in the previous case: $B \cdot \leftrightarrow \cdot N$. (This line of reasoning is valid because the dimensions of the cores of N, B and C are not too different). One could remark that the polarity of the bond is opposite to the direction of the arrow traditionally indicating the coordinative bond. It is not necessary to suspect a contradiction here; the traditional arrow only indicates the direction of a transfer of electrons in the course of a reaction, and not the eventual polarity of that bond.

ELECTRON-DEFICIENT MOLECULES; MOLECULES WITH EXCESS ELECTRONS;
LOCALIZED AND DE-LOCALIZED BONDS

Let us examine the structure of the diborane molecule, $H_2BH_2BH_2$. Figure 20 gives the geometrical arrangement of the nuclei. They are located in two perpendicular planes; the string BHBH apparently forms a square in the vertical plane. Diborane contains 16 electrons. Taking into account the boron cores with one pair of electrons with opposite spins each, a dozen electrons remains available for the bonding loges. We confirm the presence of eight pairs of neighbouring cores, and so we have $N=12$, $P=8$ and $N<2P$. This leads us to the statement that the molecule has a deficiency of electrons. In other words, it is impossible to expect the presence of a two-electron loge between every pair of neighbouring cores. The experiment indicated that the outmost B—H-bonds behave like normal single bonds. They must, therefore, be associated with two-electron loges. Four electrons are left to the four pairs constituting the BHBH-square. An *a priori* thought would lead us to a division into four one-electron loges. But the concept of loge requires specifications regarding spin. Given the usual zero-spin of ground states, the central zone should have two α-spins and two β-spins most of the time. It would then be justified to hesitate to choose between the two divisions into loges as given in Figure 21. For symmetry reasons they have the same probability P, and so $2P \leqslant 1$, and thus $P \leqslant 0.5$. A probability of this magnitude is too low to allow a good division into loges. There is no other solution to improve this probability than an extension of the loges to more than two cores. For example, the division shown in Figure 22 is unique. It has no symmetry and so it cannot be denied being a good division into loges. However, it contains two two-electron loges comprising three

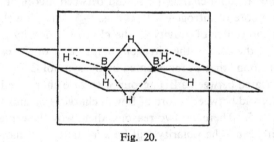

Fig. 20.

cores each. Such loges correspond with what has become known as a *de-localized bond*. We see now that a deficiency of electrons can lead to the necessity of introducing *n*-electron loges encompassing more than two cores, and demonstrating in this way the existence of delocalized bonds.

The same phenomenon often occurs in molecules where N exceeds $2P$; such molecules may be labeled electron-excessive molecules. Such is the case in benzene, where $N=30$ and $2P=24$. The analysis of this example leads to the introduction of a delocalized six-electron bond extending over six cores as shown in Figure 23, where for reasons of simplicity the two-electron loges have been replaced by the traditional lines indicating single bonds.

7. The Basis of Stereochemistry

SINGLE BONDS AND THE CONCEPT OF THE LONE PAIR

Methane provides the simplest example of a saturated hydrocarbon. In this case we have $N=2P$, and so one two-electron loge must be present between each pair of cores (Figure 24). Very often, therefore, eight electrons are found around the carbon core: four of them with spin $+\frac{1}{2}h/2\pi$, the other four with spin $-\frac{1}{2}h/2\pi$. The electrons with parallel spins must have a tendency to form angles of 109°28′ between themselves and the carbon nuclei. There will be a

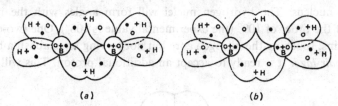

(a)   (b)

Fig. 21.

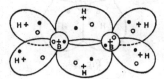

Fig. 22.

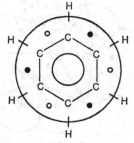

Fig. 23.

probability of finding a pair of electrons with opposite spins in the vicinity of each hydrogen nucleus; as these pairs tend to form angles of 109°28′, so will the nuclei. This accounts for the tetrahedral structure of methane. Let us now look at trimethyl-boron, $B(CH_3)_3$. It contains 32 electrons. At any given time eight of them very probably occupy the core loges. That leaves $N=24$. One can verify that $P=12$. Again, all bonds are two-electron loges. Three of these emanate from the boron, the three B—C-bonds. Most often, six electrons will move around the boron core, three with α- and three with β-spin (Figure 25). Extending the foregoing argument, one might deduce that the three carbon nuclei form angles of 120° with the boron nucleus and will therefore be co-planar with the latter. This is indeed what the experiment revals. By comparison we might similarly have thought that the three N—H-bonds in NH_3 also form angles of 120°. That is, however, not true. A more thorough analysis tells us why not. That molecule contains ten electrons and so $N=8$, i.e. four pairs of electrons with opposite spins have a high probability of being found near the nitrogen core (Figure 26). The angles between these pairs will be of the order of 109°. There is place for only two electrons with opposite spins in the vicinity of each hydrogen nucleus. The hydrogen nuclei will form angles with the nitrogen nucleus of the order of 109° (the experimental value is 108°). One two-electron loge remains, touching the nitrogen core but not serving to establish a bond: it is a *lone pair loge*. It is ready to accept an additional proton. This will lead to

Fig. 24.

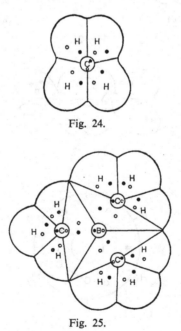

Fig. 25.

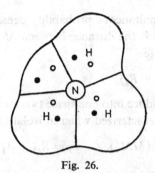

Fig. 26.

the NH_4^+ ion. Since the latter is iso-electronic with NH_3 it may be expected that the angles in NH_3 and NH_4^+ will be similar. In fact, we know that NH_4^+ is perfectly tetrahedral and so the angles are 109°28'. The conclusion to be drawn from this analysis may be that the most probable angles between single bonds emanating from the same cores of elements of the first period, are of the order of 109°, 120° and 180°. A statistical analysis of the valence angles confirms that those values occur most frequently for this kind of atom [19].

8. Intermolecular Forces

THE ENERGY OF A SYSTEM OF ELECTRONS AND THE CONCEPT OF THE SUPERMOLECULE

Consider an atom or a molecule in the approximation that its protons have been replaced by fixed nuclei. It can be shown that the virial theorem [29], mentioned before in Section 3, preserves its validity in the present case so that the total energy shall be $E = \frac{1}{2}F$. In other words, it suffices to analyse the average potential energy in order to obtain a value for the total energy. The potential energy contains three factors:

(a) The mutual repulsion of the nuclei

$$R_n = \sum_{P<Q} (Z_P Z_Q / r_{PQ}) e^2,$$

where r_{PQ} is the distance between nucleus P with charge $Z_P e$ and nucleus Q with charge $Z_Q e$;

(b) The attraction between electrons and nuclei

$$A = \sum_P (Z_P e^2 \varrho(M) / r_{MP}) dv_M,$$

where $\varrho(M)$ is the probability density of an electron at M, and r_{MP} the distance between point M and nucleus P;

(c) The interelectronic repulsion

$$R_e = \int (e^2 D(M, M') / r_{MM'}) dv_M dv_{M'},$$

where $D(MM')$ is the simultaneous probability density of one electron in M and the other in M'; $r_{MM'}$ is the distance between M and M'.

We can now simply write

$$E = \tfrac{1}{2}(R_n + A + R_e).$$

Suppose now that we introduce into one space two atomic or molecular systems which could have been characterized when in isolation, by respectively

$$\varrho_1(M), \qquad D_1(MM'), \qquad E_1 = R_{n1} + A_1 + R_{e1}$$

and

$$\varrho_2(M), \qquad D_2(MM'), \qquad E_2 = R_{n2} + A_2 + R_{e2}.$$

Evidently, the electrons in one system repel those in the other, the same holds for the nuclei, and the electrons in one system are attracted by the nuclei in the other. Instead of the functions $\varrho_1(M)$ and $\varrho_2(M)$, we must have $\varrho(M)$, and a new function $D(MM')$ instead of $D_1(M')$ and $D_2(M')$. The summations will encompass the entire set of nuclei in the two systems. A new energy E arises now for this new ensemble that could be called a *supermolecule*. Assuming, to begin with, that the two systems are sufficiently far apart to preserve their individucal characters, the energy difference

$$\Delta E = E - E_1 + E_2.$$

It is customary to decompose this energy into three terms. The so-called *electrostatic energy* is that part of E that would have existed had the mutual approach of the two systems perturbed neither the densities ϱ_1 and ϱ_2, nor the densities D_1 and D_2. The *polarization energy* represents the contribution to the energy arising from the replacement at their previous positions of the densities ϱ_1 and ϱ_2 by the real density ϱ, disregarding the way in which the motion of the electrons in one system depends on the motions of the electrons in the other. The *dispersion energy*, finally, consists of the correction introduced by taking into account the true form of the function $D(MM')$, i.e. specifically to which extent the presence of an electron at point M in one system modifies the probability density at point M' in the other.

LONG-RANGE INTERMOLECULAR FORCES

The electrostatic theory provides the proof of the possibility to replace a continuous charge by a set of point charges, dipole moments (equivalent with two equal but opposite charges on a short distance apart), quadrupole moments, ..., etcetera, which at large distances create the same field as the continuous charge would have done. The electrostatic energy between two atoms or molecular systems will therefore contain interactions of the following kinds: charge-charge, charge-dipole moment, dipole moment-dipole moment, ...

The electrostatic interaction between the four charges in Figure 27, simulating two neutral dipole molecules, may be approximated by the formula

$$E_{el} = (\mu_a \mu_b / r_{ab}^2)(-2\cos\theta_a \cos\theta_b + \sin\theta_a \sin\theta_b)$$

with $\mu_a = e_a d$ and $\mu_b = e_b d$. The four charges have been assumed coplanar. The formula gives the energy correct to within 10%, provided that the distance

Fig. 27.

r_{ab} is over five times as large as the larger of d and d'. The origin of the polarization energy can easily be understood in a simple case. For example, move a hydrogen atom in its ground state towards a proton. Before the proton's approach the hydrogen's density $\varrho(M)$ has a spherical symmetry. The approaching proton makes that this symmetry disappears. Exerting an attractive force on the electron, its effect is a tendency towards a larger ϱ at points M close to the proton, and towards a smaller ϱ at more distant points. The proton is said to polarize the atom. By and large, this phenomenon may be accounted for by the assumption that under the influence of an electric field $\mathbf{h}$ a molecule acquires an induced dipole moment $\mathbf{\mu} = \alpha\mathbf{h}$ with α a characteristic property of the molecule, the polarizability. When two neutral molecules may be represented as in Figure 27, by two dipoles, the polarization energy shall be given by the formula

$$E_{pol} = -[\mu_a^2 \alpha_b (3\cos^2\theta_a + 1) + \mu_b^2 \alpha_a (3\cos^2\theta_b + 1)]/2r^6,$$

where α_a and α_b are the respective polarizabilities of the two molecules. In order to clearly demonstrate the origin of the dispersion forces, we choose the example of two hydrogen atoms with a large inter-atomic distance. Having zero charge and carrying no dipole moment, the atoms exercise neither electrostatic interaction nor polarization effects. Let us consider (Figure 28) the case

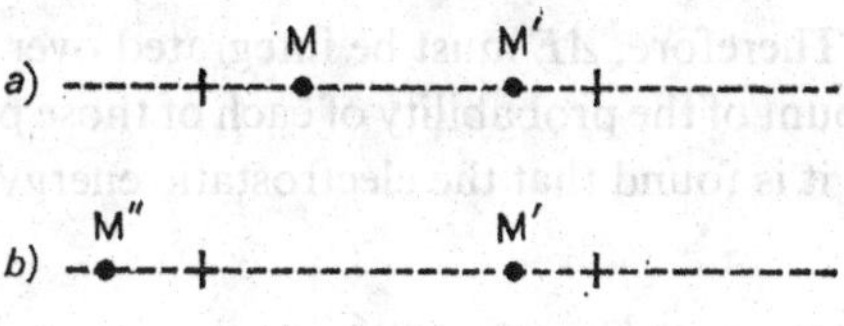

a)   M   M'

b)   M"   M'

Fig. 28.

of two atoms with a large initial distance. Be $\varrho(M)$ the electronic density at M in one atom, and $\varrho(M')$ the electronic density at M' in the other; the simultaneous density $D(MM')$ shall then be

$$D(MM') = \varrho(M)\,\varrho(M')$$

since the probability of the simultaneous occurrence of two independent events is simply the product of the probabilities of the isolated events. We direct our attention now to a point M'' (Figure 28b), a symmetrical counterpart of M with respect to the nucleus of the first-mentioned atom. Then

$$\varrho(M'') = \varrho(M),$$

and so

$$D(M'', M') = \varrho(M'')\,\varrho(M') = \varrho(M)\,\varrho(M') = D(MM').$$

In other words, the events a and b are equally probable. Now we allow the atoms to come closer together so that they no longer can disregard each other. Then of course,

$$D(MM') \neq \varrho(M)\,\varrho(M')$$

and we intuitively expect event (a) to become less probable than event (b), since in the first case the electrons are closer and for that reason repel each other more strongly than in case (b), the latter being energetically less favourable. The modification of $D(MM')$ gives rise to the so-called dispersion forces. One might say that the momentary dipole moment appearing when an electron is at M' in the second atom, favours the appearance of a momentary dipole moment linked to the presence of an electron at M'' in the first atom. It can be shown that in simple cases the dispersion energy can conveniently be given by the formula

$$E_{\text{disp}} = -\tfrac{3}{2}(\alpha_a\alpha_b/r_{ab}^6)(I_aI_b/I_a + I_b),$$

where I_a and I_b are the ionization energies of the actual systems. For two neutral molecules the intermolecular energy at a large distance is obtained from the summation

$$\Delta E = E_{\text{el}} + E_{\text{pol}} + E_{\text{disp}}.$$

The sum depends on the relative orientations of the molecules, in particular on the angles θ_a and θ_b. Therefore, ΔE must be integrated over all possible orientations, giving due account of the probability of each of those possible orientations. After the integration it is found that the electrostatic energy varies with r_{ab}^{-6} and ΔE is given by

$$\Delta E = -(K_{\text{el}} + K_{\text{pol}} + K_{\text{disp}})/r_{ab}^6.$$

Table IV gives the magnitudes of the coefficients K of some simple molecules. It is hard to make an *a priori* estimate as to which force is predominant. In the case of water it is the electrostatic force. For the other two molecules the dispersion force is predominant.

TABLE IV

Interacting molecules	Magnitude of K (in 10^{60} erg cm^{-6})		
	K_{el}	K_{pol}	K_{disp}
CO and CO	0.003	0.06	67
HI and HI	0.35	1.7	382
H_2O and H_2O	190	10	47

SHORT-RANGE INTERMOLECULAR FORCES

When atomic or molecular systems are at such short distances that the electronic domains overlap, the interactions are quite different. In some cases bonds are formed between the two systems. In other words, a loge is constituted between the core of one system and the core of the other. In other cases, however, the intermolecular forces will be repulsive instead of attractive as they were at large distances. Again, the example of two hydrogen atoms will lead us towards an understanding of this phenomenon. When the two electrons have opposite spins they can be together in one small domain; a two-electron loge will be formed between the two protons. When two electrons have parallel spins they cannot together occupy the same domain. This will result in a fierce repulsion once the internuclear distance has decreased to below a certain value. This is commonly referred to as *Pauli's exclusion repulsion*. The curves in Figure 29

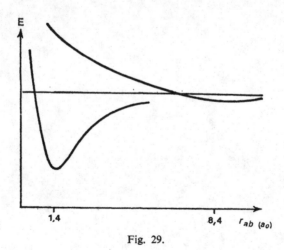

Fig. 29.

exhibit the variation of the system's energy with r_{ab} for both cases. In each case the curve has a minimum. In the first one it is located at 1.4a, with a steep trough and it represents the formation of the H_2 molecule. In the second case the minimum is found at $r_{ab} = 8.4a$ and the trough is barely noticeable; in fact, it is so shallow that it usually does not show in the diagrams. The two hydrogen atoms are weakly associated. Each constitutes one one-electron loge. Simple thermal motion can destroy the weak association. The repulsive part of the curve $E = f(r_{ab})$ is in that case rather well given by the expression

$$E = 350 \, (r_{ab}/a_0)^{-4/3} \exp - (0.357 r_{ab}/a_0) \text{ kcal mole}^{-1}.$$

Recalling what we learned about long- and short-range forces, it is now easy to understand why in simple cases the intermolecular energy is often expressed by a formula of the type

$$E = \alpha \exp (-\beta r) - \gamma/r_{ab}^6,$$

where α, β, and γ are experimentally determined for each particular case [30]. The existence of the Pauli exclusion forces allows us to understand the origin of the repulsion known as *steric hindering*, present in a molecule when for geometrical reasons non-bonded atoms are squeezed together.

CHAPTER II

SURVEY OF COMPUTATIONAL TECHNIQUES

9. The Independent-Electron Model; the Non-Invariant Concept of Orbital

THE NON-INDEPENDENCY OF THE ELECTRONS IN THE INDEPENDENT-ELECTRON MODEL

In his book Le Rationalisme Appliqué, Bachelard [3] suggests to add a rule of explicit exorcism of false ideas to the rule of enumeration of true ideas. *A priori* one might expect this rule to be of little use in the very operational domain of computational techniques which we are about to enter. In reality, the situation is diametrically opposite. This branch of science abounds in false notions. In my opinion this may be understood by recognizing the existence of two extreme classes of researchers: the veritable theoretician who avoids attaching a physical meaning to even the humblest of his technical operations, and the more hybrid workers who, lacking in an adequate mathematical training, endeavour to attach delusions with illusory explicative powers to mathematical subjects, the true meaning of which they fail to understand. It must be admitted that the specialist's language is particularly poorly designed, as though composed to suggest false notions. I will try not to overlook the application of the rule of exorcism, whenever necessary.

An analysis of the independent-electron model will provide me with the first opportunity. The model in question constitutes the essential technical basis of the majority of computational techniques. Let us go back to the example of the helium atom of Section 4. The wave function may be written

$$H\Psi = (T_1 + T_2 + F_1 + F_2 + e^2/r_{M_1 M_2})\,\Psi = E\Psi$$

with

$$T_1 = -(h^2/8\pi^2 m)\,\Delta_1; \; T_2 = -(h^2/8\pi^2 m)\,\Delta_2; \; F_1 = -(2e^2/r_{M_1});$$
$$F_2 = -(2e^2/r_{M_2}).$$

The problem with the solution of this equation is the presence of the term $e^2/r_{M_1 M_2}$, which represents the Coulomb repulsion between the two electrons. The independent-electron model amounts to the neglect of this operator. That leads to solving the equation

$$H^0\Phi = (T_1 + T_2 + F_1 + F_2)\,\Phi = E^0\Phi$$

with $H^0 = H - (e^2/r_{M_1 M_2})$. Examine now the equation

$$(-(h^2/8\pi^2 m)\,\Delta - (2e^2/r_M))\,\varphi_i(M) = \varepsilon_i \varphi_i(M). \tag{2.1}$$

It represents the movements of an isolated electron in the field of a fixed helium nucleus. It can easily be understood that a product $\varphi_i(M_1)\,\varphi_j(M_2)$ is an exact solution of the independent-electron model. In fact

$$(T_1 + T_2 + F_1 + F_2)\,\varphi_i(M_1)\,\varphi_j(M_2) =$$
$$= \varphi_j(M_2)\,(T_1 + F_1)\,\varphi_i(M_1) + \varphi_i(M_1)\,(T_2 + F_2)\,\varphi_i(M_2)$$
$$= \varphi_j(M_2)\,\varepsilon_i\varphi_i(M_1) + \varphi_i(M_1)\,\varepsilon_j\varphi_j(M_2) =$$
$$= (\varepsilon_i + \varepsilon_j)\,\varphi_i(M_1)\,\varphi_j(M_2).$$

The function $\varphi_i(M_j)\,\varphi_j(M_2)$ is apparently an eigenfunction of the operator H^0 for the eigenvalue $E^0 = \varepsilon_i + \varepsilon_j$.

The functions φ are called the atomic orbitals of helium.

The energy of the atom is thus simply reduced to the sum of the energies associated with the orbitals present in a wave function belonging to the independent-electron model. The solutions of Equation (2.1) are quite similar to those of the equation of the hydrogen atom; they are hydrogen-like functions defined by the same quantum numbers. One may still speak of the orbitals in terms of shell, state, etc. The K-orbital is now written

$$\varphi_K(M) = N \exp - (2r_M/a_0).$$

The ground state of the helium atom will be represented by the wave function

$$\Phi_{KK}^0(M_1 M_2) = \varphi_k(M_1)\,\varphi_k(M_2) =$$
$$= N^2 \exp - (2r_{M_1}/a_0) \exp - (2r_{M_2}/a_0)$$

in the approximation of the independent-electron model, and the energy of the atom becomes

$$E_{KK}^0 = 2\varepsilon_K = -108 \text{ eV}.$$

However, the experimental value is -78.6 eV, and this reveals that the neglect of the repulsion between the two electrons leads to a stabilization of the atom by about 30 eV. The model under examination is apparently too crude from a quantitative point of view, but we will see that it is a good starting point for better approximations. In view of this we will continue the analysis. We did already recall one of the fundamental principles of probability theory. Two events are called independent when the probability of simultaneous occurrence P equals the product of the probabilities of the isolated events. The simultaneous probability density $D(M_1 M_2)$ of electron 1 in M_1 and 2 in M_2, is here

$$D(M_1 M_2) = |\Phi_{KK}^0|^2 = |\varphi_K(M_1)|^2 |\varphi_K(M_2)|^2.$$

The probability density $\varrho(M_1)$ of electron 1 at M_1 is

$$\varrho(M_1) = |\varphi_K(M_1)|^2 \quad \text{and likewise} \quad \varrho(M_2) = |\varphi_K(M_2)|^2,$$

and so

$$D(M_1 M_2) = \varrho(M_1)\, \varrho(M_2).$$

So the movements of the electrons are independent and we are tempted to say that each electron is guided by a wave φ_K, that there are two electrons in the orbital φ_K. In the general case of a state represented by

$$\Phi_{ij}^0 = \varphi_i(M_1)\, \varphi_j(M_2)$$

the wave φ_i is said to guide one electron and the wave φ_j the other, or that one electron is in orbital φ_i and the other in orbital φ_j. It is only one step from this point to assigning the quantum numbers of orbital φ_i to one electron and those of φ_j to the other. This step is even made easier by considering the fact that the energy of state Φ_{ij} equals $E_{ij} = \varepsilon_i + \varepsilon_j$, and that for example, by comparing Φ_{KL} having $E_{KL} = \varepsilon_K + \varepsilon_L$ with Φ_{KK} having $E_{KK} = 2\varepsilon_K$, it could be maintained that one electron has been excited from the K-orbital to the L-orbital. This is the dangerous language of the shell-model. Time has come to exorcise such usage as we are about to arrive at a state of conflict with the principle of indistinguishability. It is easy to see the roots of the myth. Quite simply, the wave functions should have been properly *anti-symmetrized*. In actual fact, a function such as

$$\varphi_K(M_1)\, \varphi_L(M_2)$$

is unacceptable; no more symmetrical than anti-symmetrical, it can neither represent nor even approximate a state of helium. It has to be symmetrized, and the choice must be either

$$\Phi_{KL}^S = \varphi_K(M_1)\, \varphi_L(M_2) + \varphi_L(M_1)\, \varphi_K(M_2)$$

or

$$\Phi_{KL}^A = \varphi_K(M_1)\, \varphi_L(M_2) - \varphi_L(M_1)\, \varphi_K(M_2),$$

each one of these being a proper eigenfunction of H° for the energy

$$E_{KL}^0 = \varepsilon_K + \varepsilon_L \quad \text{since} \quad \varepsilon_L + \varepsilon_K = \varepsilon_K + \varepsilon_L.$$

The functions $\Phi_{KL}^S \sigma^A(\sigma_1, \sigma_2)$ and $\Phi_{KL}^S \sigma^S(\sigma_1, \sigma_2)$ obey the Pauli principle and truly are functions in the independent-electron model. With a function like

$$\Phi_{ij} = \varphi_i(M_1)\, \varphi_j(M_2) \pm \varphi_j(M_2)\, \varphi_i(M_1)$$

it is easy to see that except for $i = j$, then $D(M_1 M_2) \neq \varrho(M_1)\, \varrho(M_2)$. The proof is left to the reader. The electrons are not independent in the independent-electron model. The independent-electron model introduces a strong correlation between the movements of the electrons. And luckily so, as otherwise that model would have been catastrophically bad considering that two electrons with parallel spins cannot be in the same small domain; the entering electron would chase away the electron already present. We know that this is a fundamental phenomenon which plays an essential part in connection with the geometrical conformation of the molecule. Should the independent-electron model have failed to account for that, it would have been useless. Fortunately, such is not the case. Let us reject the language of the shell model which suggests so many false notions, and rather speak of the pseudo-independent-electron model. This practice will bar us from jumping too hastily to conclusions. An electron shall not be assigned to any particular orbital. Orbitals shall be used for the construction of approximations to the wave functions representing the collective behaviour of the electrons present in the system. There is no one-to-one correspondence between electrons and orbitals. All orbitals play their roles simultaneously in determining an electron's whereabouts. This may be shown with the function

$$\Phi(M_1, M_2) = 2^{-1/2}(\varphi_K(M_1)\,\varphi_L(M_2) + \varphi_L(M_1)\,\varphi_K(M_2)).$$

Allowing of the simplification that the functions φ are real functions, we find then

$$D(M_1 M_2) = |\Phi(M_1, M_2)|^2 = \tfrac{1}{2}\varphi_K(M_1)^2\,\varphi_L(M_2)^2 + \\ + \tfrac{1}{2}\varphi_L(M_1)^2\,\varphi_K(M_2)^2 + \varphi_K(M_1)\,\varphi_L(M_1)\,\varphi_K(M_2)\,\varphi_L(M_2).$$

The probability density of electron 1 in M_1 is then

$$\varrho(M_1) = \int D(M_1 M_2)\,dv_{M_2} = \tfrac{1}{2}(\varphi_K^2(M_1) + \varphi_L^2(M_1))$$

by virtue of the orthogonality of the eigenfunctions φ_K and φ_L:

$$\int \varphi_K(M_2)\,\varphi_L(M_2)\,dv_2 = 0.$$

THE NON-INVARIANT NATURE OF THE CONCEPT OF ORBITAL

In order to continue with our exorcising work we may say that there is no *a priori* correspondence between properties of orbitals and those of electrons. The total wave function describes the movements of the electrons. When we wish to understand the structure of atoms and molecules, we must invoke its mediation. The problem of 'imagining' a total wave function has lead certain workers to the establishment of erroneous links between electrons and orbitals. Through a demonstration of the resulting inconsistencies we hope to be able to provide a better understanding of the fundamental principles determining the behaviour of a system of electrons. We like to emphasize the importance of the concept of orbital as an extremely powerful tool. The problem is to put it to a proper use without walking straight into the many traps it sets out for us. It is essential to be perfectly aware of the existence of an infinite number of sets of orbitals exactly defining the same wave function.

The function

$$\Phi^A_{KL} = \varphi_K(M_1)\,\varphi_L(M_2) - \varphi_L(M_1)\,\varphi_K(M_2)$$

which appeared before in connection with the helium atom, can be written as a determinant

$$\Phi^A_{KL} = \begin{vmatrix} \varphi_K(M_1) & \varphi_L(M_1) \\ \varphi_K(M_2) & \varphi_L(M_2) \end{vmatrix} = \det \varphi_K(M_1)\,\varphi_L(M_2).$$

(In the last term's shorthand notation the string of symbols following 'det' are the diagonal elements only.)

Elements in rows and columns may be replaced by linear combinations of the elements in those rows and columns without thereby changing the value of the determinant. Taking for instance

$$\xi_1 = a\varphi_K + b\varphi_L \quad \text{and} \quad \xi_2 = a'\varphi_K + b'\varphi_L$$

and choosing the coefficients so as to have them obey the relation $ab' - a'b = 1$, we find

$$\det \varphi_K(M_1)\,\varphi_L(M_2) = \begin{vmatrix} \xi_1(M_1) & \xi_2(M_1) \\ \xi_1(M_2) & \xi(M_2) \end{vmatrix} = \det \xi_1(M_1)\,\xi_2(M_2).$$

The transition from φ to ξ is a so-called *unitary transformation*. An infinite number of sets of functions ξ is capable of representing the same wave function $\Phi^A_{KL}(M_1, M_2)$ leaving unchanged the latter's value anywhere in space as well as all of its properties. There is a set of such functions for every set of numbers a, a', b, b' obeying the relation $ab' - a'b = 1$. It has become common practice to generalize the notion of orbital to linear combinations like $\xi = a\varphi_K + b\varphi_L$. Whereas a and b are simultaneously non-zero, it is called a *hybrid orbital*. With this convention regarding usage, an infinite number of sets of orbitals may represent the same function, the same state of the atom. Without introducing any change in the wave functions, atomic orbitals may or may not be hybridized from the beginning. The representation of wave functions on a basis of orbitals and the projection of vectors on the coordinate axes are similar operations. It is well-known that in the latter case the lengths of the projections of a vector change upon rotation of the axes.

The function Φ^A_{KL} represents the triplet state of the helium atom which we encountered while studying the theory of loges. Should we be looking for a direct bridge between the two approaches and trying to assign electrons to orbitals, the notation $\det \varphi_K \varphi_L$ would suggest the presence of one electron in φ_K and one in φ_L. The functions φ_K and φ_L are localized in distinct regions of space, φ_K predominantly within a sphere S around the nucleus, and φ_L outside of it (see also Chapter III). One is tempted by this description to believe that it is possible to 'observe' the localized electron in its sphere S; had we selected, on the other hand, the functions

$$\xi_1 = 2^{-1/2}(\varphi_K + \varphi_L) \quad \text{and} \quad \xi_2 = 2^{-1/2}(\varphi_K - \varphi_L)$$

the total wave function would have become

$$\det \xi_1 \xi_2 = \det \varphi_K \varphi_L.$$

This is not a real change; however, the orbitals ξ are delocalized, they extend over the same region of space and both have a meaning outside as well as inside the beforementioned sphere. Continuing with the association of electrons and orbitals, one electron should now be 'observed' in ξ_1 and the other in ξ_2.

The two pictures given above are contradictory, and this proves the danger of establishing direct links between electrons and orbitals. It is dangerous to rule out the requirement that, as stated before, the total wave function should be used as an intermediary.

The theory of loges showed us the high probability of finding one electron in loge K and the other in loge L. In this sense the electrons exhibit some degree of localization. One is free to choose to represent localized electrons by either localized or de-localized orbitals. However, when among the infinite number of sets of orbitals there exists a set of highly localized orbitals, there will also exist very good loges, i.e. localization of the electrons. This can be demonstrated for the extreme case that φ_K equals zero outside of sphere S, and φ_L equals zero on the inside. If the φ's are real functions one may write

$$(\Phi_{KL}^A)^2 = \tfrac{1}{2}\varphi_K^2(M_1)\,\varphi_L^2(M_2) + \tfrac{1}{2}\varphi_L^2(M_1)\,\varphi_K^2(M_2) +$$
$$- \varphi_K(M_1)\,\varphi_L(M_1)\,\varphi_K(M_2)\,\varphi_L(M_2),$$

but the double product equals zero since at any point either φ_K or φ_L equals zero. The factor $\tfrac{1}{2}$ was introduced to ensure that Φ_{KL}^A be properly normalized as the orbitals themselves had already been normalized themselves. The probability to find electron 1 inside S and electron 2 outside of it is then

$$P_{12} = \int\limits_S \mathrm{d}v_1 \int\limits_{R^3-S} |\Phi_{KL}^A|^2 \,\mathrm{d}v_2 = \tfrac{1}{2} \int\limits_S \varphi_K^2(M_1)\,\mathrm{d}v_1 \int\limits_{R^3-S} \varphi_L^2(M_2)\,\mathrm{d}v_2 +$$
$$+ \tfrac{1}{2} \int\limits_S \varphi_L^2(M_1)\,\mathrm{d}v_1 \int\limits_{R^3-S} \varphi_K^2(M_2)\,\mathrm{d}v_2,$$

where R^3 stands for 'three-dimensional space'.

The first term obviously equals $\tfrac{1}{2}$ as a result of the normalization of the orbitals. The second term also equals zero. The conclusion obviously has to be that $P_{12} = \tfrac{1}{2}$. The probability P of finding one and only one electron within the sphere, be it electron 1 or electron 2, and the other outside of it, therefore equals $P_{12} + P_{21} = 1$. This means that one has the certainty that one and only one electron will be found within S. The sphere S is therefore a perfect K-loge and the remainder of space an equally perfect L-loge.

Let us now study a larger atom, such as for instance a carbon atom in the state called ^5S by spectroscopists. In order to simplify matters we will use a model in which a nucleus with a charge $6e - 4e = 2e$ takes the place of the K-shell and its surrounding four electrons. In the context of the independent-electron model the lowest ^5S state will be represented by the function

$$\Phi(M_1, M_2, M_3, M_4) = \det Ls(M_1) Lp_0(M_2) Lp_{+1}(M_3) \times$$
$$\times Lp_{-1}(M_4),$$

where

$$\varphi_{Ls} = Ls = R_{2,0}(r) \qquad \varphi_{Lp+1} = Lp_{+1} = 2^{-1/2}R_{2,1}(r)\sin\theta\exp - i\varphi$$
$$\varphi_{Lp0} = Lp_0 = R_{2,1}(r)\cos\theta \qquad \varphi_{Lp-1} = Lp_{-1} = 2^{-1/2}R_{2,1}(r)\sin\theta\exp i\varphi$$

since the hydrogen-like functions of Section 3 can be used. Should one wish to have again a direct bridge between electron and orbital, one could say that the present model contains on s-electron with spherical symmetry, one p_0-electron localized along the z-axis, and two more electrons (p_{+1} and p_{-1}) close to the Oxy-plane. The orbitals, though, may be replaced by hybrids:

$$\xi_1 = aLs + bLp_0 + cLp_{+1} + dLp_{-1}$$
$$\xi_2 = a'Ls + b'Lp_0 + c'Lp_{+1} + d'Lp_{-1}$$
$$\xi_3 = a''Ls + b''Lp_0 = c''Lp_{+1} + d''Lp_{-1}$$
$$\xi_4 = a'''Ls + b'''Lp_0 + c'''Lp_{+1} + d'''Lp_{-1}$$

under the condition that we go from the L's to the ξ's via a unitary transformation. We have now

$$\Phi(M_1, M_2, M_3, M_4) = \det Ls Lp_0 Lp_{+1} Lp_{-1} = \det \xi_1 \xi_2 \xi_3 \xi_4.$$

With a special choice of the coefficients the polarogram (= the presentation of the hybrid in polar coordinates) for a given value of r will look like the one in Figure 30, and the axes on which the ξ's have their angular maxima will make angles of $109°28'$ between themselves. One could say that the hybrids 'point towards' the corners of a regular tetrahedron centered on the nucleus, or that tetrahedral hybrids, te_1, te_2, te_3 and te_4, have been constructed. Writing the

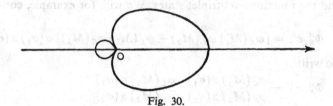

Fig. 30.

wave functions in terms of these hybrids suggests to consider that the atoms contain four electrons distributed in a tetrahedral fashion. Obviously, the first as well as the second interpretation are absolutely false. The probability density $\varrho(m)$ in a point M is

$$\varrho(M) = |Ls(M)|^2 + |Lp_0(M)|^2 + Lp_{+1}(M)|^2 + |Lp_{-1}(M)|^2 =$$
$$= |\xi_1(M)|^2 + |\xi_2(M)|^2 + |\xi_3(M)|^2 + |\xi_4(M)|^2.$$

This is definitely independent of the choice of orbitals. One can see that

$$\varrho(M) = R_{2,0}^2(r) + R_{2,1}^2(r)\left[\cos^2\theta + \sin^2\theta\right] = R_{2,0}^2(r) + R_{2,1}^2(r).$$

The electron density of carbon in the ^5S-state is independent of the angles. It represents spherical symmetry.

THE CONCEPT OF SPIN ORBITAL AND THE CONSEQUENCES OF THE PAULI PRINCIPLE FOR ITS APPLICATION

In section five it was noted that a two-electron system can be represented by either a function $\Phi^A\sigma^S$ or a function $\Phi^S\sigma^A$. It is convenient to develop the spin functions on a basis of two functions. The function $\alpha(\sigma)$ is defined by the relations

$$\alpha(+\tfrac{1}{2}h/2\pi) = 1 \quad \text{and} \quad \alpha(-\tfrac{1}{2}h/2\pi) = 0.$$

The similar function $\beta(\sigma)$ is defined by

$$\beta(+\tfrac{1}{2}h/2\pi) = 0 \quad \text{and} \quad \beta(-\tfrac{1}{2}h/2\pi) = 1.$$

The anti-symmetric spin function will be then

$$\sigma^A = \alpha(\sigma_1)\beta(\sigma_2) - \beta(\sigma_1)\alpha(\sigma_2)$$

and the three symmetric functions

$$\sigma_0^S = \alpha(\sigma_1)\beta(\sigma_2) + \beta(\sigma_1)\alpha(\sigma_2)$$
$$\sigma_+^S = \alpha(\sigma_1)\alpha(\sigma_2)$$
$$\sigma_-^S = \beta(\sigma_1)\beta(\sigma_2).$$

For a function Φ^S the only possible combination is with σ^A, but a function Φ^A can be combined with three different spin functions. It is easier to understand now why the functions Φ^S were said to represent singlet states of helium, and the functions Φ^A triplet states. We will, for example, consider the function

$$\Phi_{KL}^A\sigma_+^S = \left[\varphi_K(M_1)\varphi_L(M_2) - \varphi_L(M_1)\varphi_K(M_2)\right]\alpha(\sigma_1)\alpha(\sigma_2).$$

This can be written

$$\Phi_{KL}^A\sigma_+^S = \begin{vmatrix} \varphi_K(M_1)\alpha(\sigma_1) & \varphi_L(M_1)\alpha(\sigma_1) \\ \varphi_K(M_2)\alpha(\sigma_2) & \varphi_L(M_2)\alpha(\sigma_2) \end{vmatrix}.$$

A product $\varphi\alpha$ or $\varphi\beta$ is called a *spin orbital*, and determinants written on the basis of such spin orbitals have become known as *Slater determinants*. $\Phi^A_{KL}\sigma^S_+$ is a Slater determinant. Wave functions of the independent-electron model can be shown, in general, to be linear combinations of Slater determinants. Also, a determinant vanishes when two rows or two columns are identical. Thus it appears that the model does not accept the same spin orbital twice in the same wave function. One orbital may be used once with the α-, and once with the β-function, i.e. not more frequently than twice altogether. This is the most important consequence of the anti-symmetry of the functions, or rather of the Pauli principle. Finally, we like to indicate that the function $\Phi^A_{KL}\sigma^S_+$ may be written in a different way as

$$\Phi^A_{KL}\sigma^S_+ = \sum_P (-1)^P P\varphi_K(M_1)\, \varphi_L(M_2)\, \alpha(\sigma_1)\, \alpha(\sigma_2),$$

where the first symbol is called the anti-symmetrizer, prescribing all possible permutations of the indices 1 and 2 of the four functions (without forgetting the identity permutation), providing each permutation with a + or − sign (determined by $(-1)^P$) and imposing the summation of all the products thus obtained. The importance of this formulation stems from its broad generality.

THE NON-INVARIANCE OF THE EXCHANGE ENERGY; THE DEFINITION OF THE MOST-LOCALIZED ORBITAL

The energy of the helium atom in its lowest triplet state can be written

$$E = \int [\varphi_K(M_1)\, \varphi_L(M_2) - \varphi_L(M_1)\, \varphi_K(M_2)]\, H^0\, [\varphi_K(M_1) \times$$
$$\times \varphi_L(M_2) - \varphi_L(M_1)\, \varphi_K(M_2)]\, dv_{M_1}\, dv_{M_2} =$$
$$= 2 \int \varphi_K(M_1)\, \varphi_L(M_2)\, H^0\varphi_K(M_1)\, \varphi_L(M_2)\, dv_{M_1}\, dv_{M_2} +$$
$$- 2 \int \varphi_K(M_1)\, \varphi_L(M_2)\, H^0\varphi_L(M_1)\, \varphi_K(M_2)\, dv_{M_1}\, dv_{M_2}.$$

The first term, Q in our notation, is sometimes called the *Coulomb energy* and the second, A, *exchange energy*. This allows us to write

$$E = Q + A.$$

A new myth was born out of this equation, according to which an exchange of electrons between orbitals would create energy. Lennard Jones and Pople [31] proved the absurdity of this allegation by showing that a replacement of the functions φ by hybrids ξ does not bring about a change of the total energy E (which by its very nature is invariant) but does alter the ratio

$$\frac{A}{Q} = \frac{\xi_1(M_1)\, \xi_2(M_2)\, H^0\, \xi_2(M_1)\, \xi_1(M_2)\, dv_{M_1}dv_{M_2}}{\xi_1(M_1)\, \xi_2(M_2)\, H^0\xi_1(M_1)\, \xi_2(M_2)\, dv_{M_1}dv_{M_2}}.$$

To a great extent this ratio depends on the choice of the coefficients a, a', b and b'. These authors found examples in which A/Q varied between 0.024 and 0.28. The exchange energy has no physical meaning. It arises from a non-invariant technicality partitioning the total energy. From an examination of A it follows that the larger the product $\xi_1(M)\,\xi_2(M)$ is in a considerable part of space, the greater the overlap is between them and that the exchange energy will be more predominant as they are more delocalized. In connection with this Brion and Daudel [32] suggested to call the orbitals with the smallest ratio A/Q the most-localized orbitals. Ruedenberg put this criterion to an extensive use, and so it has become known as the Ruedenberg criterion. Forster and Boys [33] on the other hand, proposed to call those orbitals most-localized which have their centers of gravity as far apart as possible. These two criteria lead to practically the same results. One can tell from intuition that the regions occupied by the most-localized orbitals, constitute loges. This was demonstrated for a special case [34].

10. Improvement and Extension of the Independent-Electron Model

VARIATION METHODS; THE CONCEPTS OF SCREENING EFFECT AND SELF-CONSISTENT FIELD

There are several approaches towards an improvement of the independent-electron model. The basis for many of those is a variation principle. Eckart's theorem provides a starting point for the formulation of such a principle. Be

$$H\Psi(M_1, M_2) = E\Psi(M_1, M_2)$$

the characteristic equation for a stationary state of a two-electron system. Consider a set of functions $G(M_1, M_2)$, each member of which has been normalized to unity and tagged with a parameter λ. We define a set

$$D = \{G_\lambda(M_1, M_2)\}$$

obeying

$$\int |G_\lambda(M_1, M_2)|^2 \, dv_{M_1} \, dv_{M_2} = 1 \,.$$

It can be shown that for any G

$$I = \int G_\lambda^* H G_\lambda \, dv_{M_1} \, dv_{M_2} \geqslant E_1 \,,$$

where E_1 is the lowest eigenvalue of H. In order to find an approximation to the eigenfunction Ψ_1 within the set D, it seems reasonable to focus one's attention on that G that makes I as close to E_1 as possible. Since I always remains larger than or equal to E_1, a function G minimizing I has to be selected. This procedure amounts to a variation principle for the choice of G.

Let us now take a following example. We saw before that the ground state of helium in the independent-electron model is given by

$$\Phi_{KK} = N \exp(-2r_1/a_0) \exp(-2r_2/a_0).$$

The model does not take into account the interelectronic repulsion. The mutual interactions of the electrons may be accounted for by replacing the helium nucleus with one having a smaller charge Ze.
The function becomes

$$G_Z = N \exp(-Zr_1/a_0) \exp(-Zr_2/a_0).$$

The best value of Z may be found by determining the minimum of

$$I = \int G_Z H G_Z \, dv_M \, dv_{M'},$$

i.e. by taking $\partial I/\partial Z = 0$. One will find $Z = 2 - \frac{5}{16}$.

Sometimes this result is explained by making the assumption that one electron acts as a screen between the nucleus and the other electron. This screening effect adopts in the present case a value of approximately 0.3 e. Slater designed some rules enabling us to construct orbitals with built-in *screening effect* for all atoms. Those are the *Slater rules*. The Z found through application of these rules is often called the *effective atomic number*. Notice that for helium the integral I equals -77 eV with $Z = 2 - \frac{5}{16}$. This makes a difference of only 1.8 eV with the experimentally determined energy. The designation orbital is often granted to the functions $\exp(-Zr/a_0)$ and to those derived from the Slater rules, although they are no longer solutions of the equation following from the independent-electron model. An even more general way of extending the concept, again improving the model, is to consider the set of functions

$$J(M_1, M_2) = N\varphi(M_1)\,\varphi(M_2),$$

where $\varphi(M_n)$ is any normalized function, and J is normalized as well. One has to find φ so that

$$I = \int J^* H J \, dv_M \, dv_{M'}.$$

is minimized. That is *Hartree's method of the self-consistent field*. It gives a value of -77.8 eV for the ground state of helium and defines the best so-called orbital φ to represent this state.

In general J will be of the form

$$J(M_1, M_2) = \sum_P (-1)^P P\varphi_a(M_1)\,\varphi_b(M_2)\,\sigma(\sigma_1, \sigma_2),$$

where σ is a convenient spin function. When such a function remains restricted to only one determinant, the method is called the *Hartree-Fock self-consistent field method*, i.e. the self-consistent field method extended to a more general case [35].

One could also use a number of trial functions such as

$$\varphi_a(M_1)\, \varphi_b(M_2)\, f(r_{12}),$$

where r_{12} is the distance between M_1 and M_2. This method of correlated orbitals yields excellent results.

The majority of these methods have been generalized to the case of n electrons. More precise descriptions of such methods may be found in references [19] and [27a].

CONFIGURATION INTERACTION

The technique of configuration interaction is a rather different way of improving the independent-electron method. Again, let us consider the equation for helium

$$H^0\Phi = E^0\Phi,$$

a solution of which is provided by an orbital product of two orbitals φ_a and φ_b. Since an infinite number of orbitals can solve

$$(T + F)\, \varphi = \varepsilon\varphi,$$

there will also exist an infinite number of products $\varphi_a\varphi_b$, and so of solutions Φ. The configuration interaction method is based on the almost exact hypothesis that the set of such solutions forms a space of normalized functions depending on two variables, and makes up a complete basis in that space. Under these conditions an exact solution of the equation $H\Psi = E\Psi$ may be expanded into a series on that basis

$$\Psi = \sum_i c_i\Phi_i.$$

For all practical purposes the series expansion is limited to a small number of terms. The basis is truncated and in most cases remains limited to the functions Φ_i belonging to the lowest eigenvalue of H^0. What remains to be done, however, is to determine the best coefficients c_i. MacDonald's theorem is related to Eckart's, and defines the following procedure. Suppose we confine ourselves to two functions Φ_i. The following two equations may be written

$$c_1\,[H_{11} - I] + c_2\,[H_{12} - I\Delta_{12}] = 0$$
$$c_1\,[H_{21} - I\Delta_{21}] + c_2\,[H_{22} - I] = 0,$$

where

$$H_{ij} = \int \Phi_i^* H \Phi_j \, dv_M \, dv_{M'}$$

$$\Delta_{ij} = \int \Phi_i^* \Phi_j \, dv_M \, dv_{M'}.$$

These are the secular equations. The unknowns are c_1, c_2 and I. The equations being linear and homogeneous in c_i (the right-hand members are zero), they have non-zero solutions only when the determinant of the terms in brackets is equal to zero. We may write

$$[H_{11} - I] [H_{22} - I] = [H_{12} - I\Delta_{12}] [H_{21} - I\Delta_{21}].$$

This so-called secular equation is of the second degree in I and has therefore in general two roots which we will classify by letting $I_1 < I_2$. MacDonald's theorem tells us that $I_1 \geqslant E_1$ and $I_2 \geqslant E_2$. Now we have to substitute the root I_1 into the secular equation in order to find the solutions c_{11} and c_{21}. The function $c_{11}\Phi_1 + c_{21}\Phi_2$ minimizes the total energy and is considered to be the best function in the set defined by

$$\sum_{i=1}^{i=2} c_i \Phi_i$$

to represent the ground state. Subsequently the root I_2 is substituted, yielding the solutions c_{12} and c_{22} in addition to the function $c_{12}\Phi_1 + c_{22}\Phi_2$ for the first excited state.

MOLECULAR ORBITALS AND THE APPROXIMATION BY A LINEAR COMBINATION OF ATOMIC ORBITALS

The majority of the arguments put forward so far may also be used when we deal with molecules. The Hamiltonian for the hydrogen molecule (Section 4) is written

$$H = T_1 + F_1 + T_2 + F_2 + e^2/r_{12}$$

with

$$T_1 = -(h^2/8\pi^2 m) \Delta_1 \qquad T_2 = -(h^2/8\pi^2 m) \Delta_2$$
$$F_1 = -(e^2/r_{M_1 A}) - (e^2/r_{M_1 B}) \qquad F_2 = -(e^2/r_{M_2 A}) - (e^2/r_{M_2 B})$$

This notation is formally the same as the one used for the helium atom, so that a product of functions φ, solutions of $(T+F)\varphi = \varepsilon\varphi$, forms a solution Φ of the equation in the independent-electron model:

$$H^\circ \Phi = E^\circ \Phi \quad \text{with} \quad H^\circ = H - (e^2/r_{12});$$

the functions φ acquired the name *molecular orbitals*.

The model may be improved on by using either one of the self-consistent methods or the method of configuration interaction. The difficulties one encounters when trying to find exact solutions of $(T+F)\,\varphi = \varepsilon\varphi$, reveal a peculiar problem. It did not exist in the case of atoms. Its origin has to be found in the presence of several nuclei. Most often we restrict ourselves to the use of an approximation to the exact form of the orbitals, the linear combinations of atomic orbitals, the nature of which we will try to clarify with the aid of the example provided by the molecular ion H_+. When the electron is close to nucleus A, a negligeable field is experienced from nucleus B; the conclusion might be that the function describing the electron in the vicinity of A should resemble the atomic wave function centered at A, say χ_A. For symmetry reasons it must be assumed that close to B the wave function resembles an atomic function centered at B, say χ_B. We write then

$$\varphi = c_A\chi_A + c_B\chi_B.$$

The calculation of the best coefficient is essentially the same as the one we encountered with the method of configuration interaction. According to MacDonald's ideas, the equation is solved and this is followed by the determination of the coefficients. In an entirely general way a molecular orbital may be expanded as a linear combination of all the atomic orbitals deemed important for the purpose:

$$\varphi = \sum_i c_i\chi_i.$$

The orbitals thus obtained are substituted into the expression for the total wave function and the c_i are calculated with the aid of a variational method [36].

TECHNICAL FOUNDATIONS OF THE THEORY OF LOGES

The orbitals introduced via the preceding computational techniques have basically a high degree of de-localization. They extend over all the atoms. The theory of loges enables us to find the localization of the electrons hidden behind this apparent delocalization.

The radii of the loges appearing in Table I (Section 5) were found by Hartree-Fock self-consistent field calculations. The wave functions used to establish the principal results regarding loges with respect to cores and bonding, depended on functions derived from de-localized orbitals. The localization brought about by the loges has not been built into the model from the beginning, but is rather a truly mathematical demonstration of the chemist's intuition and ingenuity. Incidentally, it is simpler to begin with the same orbitals and to determine the most-localized orbitals with the aid of one of the criteria indicated in the preceding sections, in order to gain insight in the best division of the molecule into loges.

11. Methods for the Calculation of Wave Functions, Assuming that Electrons can be Localized; Loge Functions, Core Functions, Bond Functions

LOGE FUNCTIONS

A division into loges can in principle only be obtained from a wave function. And yet the theory of loges may serve as a basis for the calculation of wave functions, but above all as an educational presentation of important computational methods, often discovered in the past along entirely different lines. There is a multitude of reasons for this fact. The first one is based on the possibility to improve wave functions by iteration. A starting function is selected by one of the methods set forth in Section 10. From this, a division into loges is deduced. Relying on the localizability of the electrons it will bring about, a formalism adapted to the particular features of the molecule, a made-to-measure procedure, may be designed to begin with the next better approximation. The second reason originates in empirical data. The existence of localizability is translated into the existence of certain characteristic properties of the molecule or family of molecules: the systematics of additive molecular properties, for instance. Considering the general results explained in Sections 6, 7 and 8 in connection with the theory of loges, such experimental indications may at least suggest the topology of a good division into loges. As we are going to show, such an indication is sufficient to define an adapted formalism for the calculation of wave functions. Fortunately it is unnecessary to have any *a priori* knowledge of the correct position of the boundaries of the loges. Suppose that experimental indications interpreted in terms of the theory of loges, suggests a molecule *ABCD* possesses a lone pair on *A*, a localized two-electron bonding loge between *A* and *B*, and a de-localized four-electron bonding loge around *B*, *C* and *D*, all this around the four cores reduced to one *K*-loge each. A good division into loges may be represented topologically by Figure 31. Each loge in this figure contains a number of black dots equal to the number of electrons present there

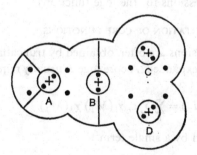

Fig. 31.

with a high probability. This topology suggests a choice of functions

$$D = \sum_P (-1)^P P K_A(M_1, M_2) K_B(M_3, M_4) K_C(M_5, M_6) \times$$
$$\times K_D(M_7, M_8) P_A(M_9, M_{10}) L_{AB}(M_{11}, M_{12}) \times$$
$$\times L_{BCD}(M_{13}, M_{14}, M_{15}, M_{16}) \sigma(\sigma_1, ..., \sigma_{16}).$$

The functions K represent the various cores; the functions P_A symbolize the lone-pair loge; the function L_{AB} represents the bonding loge between the cores A and B; the function L_{BCD} represents the de-localized bonding loge. They were introduced by the present author [19].

In order to obtain the best *loge functions* it suffices to follow the same route used to obtain the best orbitals: a search is made for the functions minimizing

$$I = \int D^* H D \, dv_{M_1} ... dv_{M_{16}}.$$

The results are far superior to those found with the self-consistent field method, this because all the correlations between the electrons in a loge are taken care of. The calculations require more time but are not at all unsolvable. For all practical purposes the unknown functions are expanded on a basis of one-center functions, such as atomic orbitals or Gauss functions. For example, one could use the following expression:

$$L_{BCD} = \sum_i \sum_j \sum_k \sum_l c_{ijkl} \chi_i(M_{13}) \chi_j(M_{14}) \chi_k(M_{15}) \chi_l(M_{16}),$$

and the variation-method gives the best value for c_{ijkl}.

In actual practice the method of the loge functions and of the self-consistent field cannot be used but for the very simplest molecules, and even then a powerful computer is a necessity. For that reason less elaborate methods were developed before the era of the computer, and even during that latter era, since it is possible to solve several important problems within the confines of very restricted calculations. The majority of such procedures amounts to the use of simplified expressions for the loge functions.

APPROXIMATE REPRESENTATION OF CORE FUNCTIONS

Such simplified expressions are often obtained by truncating the base on which the loge functions are expanded. A core loge K usually requires the introduction of a function

$$K_A(M_1, M_2) = \sum_i \sum_j c_{ij} \chi_i(M_1) \chi_j(M_2)$$

but will be represented by a single term

$$K_A(M_1, M_2) = \chi_{AK}(M_1) \chi_{AK}(M_2),$$

where χ_{AK} is the Slater orbital with the lowest energy, 1s, associated with atom A in the context of the independent-electron model of that atom. In view of this we write

$$K_A = 1s(M_1)\, 1s(M_2) = N \exp - (Zr_1/a_0) \exp - (Zr_2/a_0)$$

with Z = effective atomic number (i.e. atomic number corrected for the screening effect).

A core consisting of a K-loge and a 'complete' L-loge shall be similarly written

$$K = 1s(M_1)\, 1s(M_2)\, 2s(M_3)\, 2s(M_4)\, 2p_0(M_5)\, 2p_0(M_6) \times$$
$$\times\, 2p_{+1}(M_7)\, 2p_{+1}(M_8)\, 2p_{-1}(M_9)\, 2p_{-1}(M_{10}).$$

These functions are completely definite and will introduce no unknowns into the variation calculation. On the other hand, the values of Z in the exponents may be left undetermined and one has to rely on the variation procedure to find the best exponents.

THE REPRESENTATION OF A LOCALIZED BONDING LOGE

Within a good approximation it could be said that a localized bonding loge is the result of an amalgamation of the parts of the atom's outer loges between which a bond is formed. One must try, therefore, to construct the loge function by making combinations of atomic orbitals that are capable of 'filling' the region of space occupied by the loge. At this stage it will be most helpful to consider the spatial orientation of the hybrid orbitals. Take the example of methane. In Figure 24, Section 7, the topology of the arrangement of the loges in that molecule have been given. The functions D are

$$D = \sum_P (-1)^P\, PK(M_1, M_2)\, L_{CH_b}(M_3, M_4) \times$$
$$\times\, L_{CH_a}(M_5, M_6)\, L_{CH_c}(M_7, M_8)\, L_{CH_d}(M_9, M_{10}),$$

where K is the core loge function of the carbon and L_{CH} the various functions of the bonding loge. Each bonding loge is the result of the fusion of a part of the L-loge of carbon with the only loge of the hydrogen atom. The latter can be represented by a 1s-function of the free atom. It is tempting to proceed according to the principles of the method of the linear combination of atomic orbitals (LCAO) and to try to represent a loge L_{CH} by a linear combination of the 1s-function with a combination of carbon-L-orbitals, thus 'filling' the loge in question as completely as possible. However, we learned in Section 9, Figure 30, about the possibility of a unitary transformation turning the orbitals Ls, Lp_0 and $Lp_{\pm 1}$ into four tetrahedral hybrids. We choose to select the coefficients so as to construct the hybrids te_a, te_b, te_c and te_d pointing towards the four hydrogen nuclei, and we write for example

$$L_{CH_a}(M_1, M_2) = [\lambda 1s_a(M_1) + \mu te_a(M_1)] [\lambda 1s_a(M_2) + \mu te_a(M_2)].$$

The coefficients λ and μ will follow from the variational method. The representation of the CH-loge can be improved by putting

$$L_{CH_a}(M_1, M_2) = \lambda 1s_a(M_1) 1s_a(M_2) + \mu te_a(M_1) te_a(M_2) +$$
$$+ v[1s_a(M_1) te_a(M_2) + te_a(M_1) 1s_a(M_2)],$$

introducing the orbital products that can be constructed on the basis $\{1s_a, te_a\}$. This is the so-called group function method [37a, b]. Although simpler than the self-consistent field method, it gives a slightly better energy $(-53.48 \text{ eV}$ against $-53.44 \text{ eV})$. The time required to compute the problem on an IBM 360/75 is of the order of 40 seconds.

THE REPRESENTATION OF LONE-PAIR LOGES

Consider a molecule such as ammonia (Figure 26). Of the four hybrids te_a, te_b, te_c and te_d we will not use te_d. This latter one is perfectly suitable to represent the lone pair, as the regions where this hybrid has a high density coincide with the corresponding loge. The correct expression is

$$P_N = te_d(M_1) te_d(M_2).$$

THE REPRESENTATION OF DE-LOCALIZED BONDING LOGES; $\sigma - \pi$ SEPARATION

We select diborane (Section 6, Figure 22) as an example of a molecule with a delocalized bond. Our task is to find the representation by a wave function-$L_{B_1HB_2}(M_1, M_2)$, of one of the de-localized two-electron loges comprising the borium atoms and one of the central hydrogen atoms. In a first approximation this wave function may be constructed from the hydrogen's $1s$-orbital and the hybrid orbitals of the two borium atoms. An examination of the environment of a boron core reveals the presence of two single-bond BH-loges and two de-localized loges. It is possible to find a unitary transformation creating four hybrids, t_{1a}, t_{1b}, t_{1c} and t_{1d}, pointing towards the singly-bonded hydrogens and to the bridging hydrogens held by the delocalized bond. Let t_K be the one pointing towards the hydrogen in the loge we are examining. On symmetry grounds a hybrid t_{2c} may be chosen as emanating from the second boron core and pointing towards the same hydrogen. The expression then will be

$$L_{B_1HB_2}(M_1, M_2) = [\lambda t_{1c}(M_1) + \mu 1s(M_1) + v t_{2c}(M_1)] \times$$
$$\times [\lambda t_{1c}(M_2) + \mu 1s(M_2) + v t_{2c}(M_2)].$$

One can do better by constructing a function similar to the one used by Klessinger and McWeeny.

The next example is benzene (Figure 23). Adopt the coordinates for the definition of the atomic orbitals of each carbon atom in such a manner that they are centered on the carbon nucleus, with the x- and y-axes in the plane of the hexagon and the z-axis perpendicular to them. The localized bonding loges may be represented with the aid of the $2s$- and the $2p_{\pm 1}$ orbitals, which as was stated before, have their highest density along the axes in the plane. The orbitals obtained this way are symmetrical with respect to the plane of the molecule. They are called σ-orbitals. The six $2p_0$-orbitals, varying with $\cos\theta$ about the z-axis, are left for the representation of the de-localized loge. They are anti-symmetrical with respect to the said plane. They are called the π-orbitals. It is unfortunate that in the present case the same symbols have to be used as those adopted for atomic orbitals to distinguish between σ- and π-character. This could lead to a dangerous confusion.

A de-localized loge must now be represented by a combination of those six remaining orbitals. A convenient method of doing this is the application of linear combinations

$$\eta = a_1\chi_1 + a_2\chi_2 + a_3\chi_3 + a_4\chi_4 + a_5\chi_5 + a_6\chi_6,$$

where $\chi_i = a2p_0$ orbital.

Since the loge function $L_{C_1\cdots C_6}$ depends on six points, at least three different orbitals must be used in order to prevent the function from vanishing under the action of the anti-symmetrizer. To this end three linear combinations with different sets of coefficients are selected and

$$L = \eta_1(M_1)\,\eta_1(M_2)\,\eta_2(M_3)\,\eta_2(M_4)\,\eta_3(M_5)\,\eta_3(M_6).$$

The coefficients have to be determined via a variational procedure. One may also write

$$L = \sum_{ijklmn} c_{ijklmn}\chi_i(M_1)\,\chi_j(M_2)\,\chi_k(M_3)\,\chi_l(M_4)\,\chi_m(M_5)\,\chi_n(M_6),$$

reverting to the group-function method. This should lead to a better wave function because of the introduction by this method of a larger number of variational parameters.

EMPIRICAL AND SEMI-EMPIRICAL METHODS

One approximation has been proved quite useful; it amounts to the assumption that a small perturbation of a molecule like benzene above all affects the π-orbitals. In view of this it is attractive to replace the molecule by a model in which the six electrons move around a charge distribution instead of around the cores and the localized bonding loges. This is the π-approximation. We will treat the electrons in accordance with the independent-electron model.

The molecular orbitals will have to be solutions of an equation of the form

$$(T + F)\, \varphi = \varepsilon\varphi \,.$$

Let $T+F=h$, and use the approximation of the linear combination of atomic orbitals. This leads to an expression which is the analogue of the one encountered in the previous paragraph

$$\varphi = a_1\chi_1 + a_2\chi_2 + a_3\chi_3 + a_4\chi_4 + a_5\chi_5 + a_6\chi_6 \,.$$

It will not be difficult now, to compute the coefficients with the aid of MacDonald's theorem. Substitute

$$\alpha_{ii} = \int \chi_i^* h\chi_i \, dv \,; \qquad\qquad \beta_{ij} = \int \chi_i^* h\chi_j \, dv \quad \text{for} \quad i \neq j \,;$$

$$S_{ij} = \int \chi_i^* \chi_j \, dv \quad \text{for} \quad i \neq j \,; \qquad \int \chi_i^* \chi_i \, dv = 1 \text{ (normalization)} \,.$$

The integrals α are usually called *Coulomb integrals*, the integrals β *resonance integrals* and the integrals S *overlap integrals*. The secular equation becomes

$$\begin{vmatrix} \alpha_{11} - I & \beta_{12} - IS_{12} & \cdots\cdots & \beta_{16} - IS_{16} \\ \beta_{21} - IS_{21} & \alpha_{22} - I & \cdots\cdots & \cdots \\ \cdots & \cdots & \cdots\cdots & \cdots \\ \cdots & \cdots & \cdots\cdots & \cdots \\ \beta_{61} - IS_{61} & \beta_{62} - IS_{62} & \cdots\cdots & \alpha_{66} - IS_{66} \end{vmatrix} = 0 \,.$$

On symmetry grounds, all the α must be equal. Hückel further neglected all the S-integrals and all integrals between non-adjacent cores. Then, simply, $\alpha_{ii} = \alpha$ and because of the symmetry $\beta_{12} = \beta_{23} = \cdots = \beta$. The secular equation has now become

$$\begin{vmatrix} \alpha - I & \beta & 0 & 0 & 0 & \beta \\ \beta & \alpha - I & \beta & 0 & 0 & 0 \\ 0 & \beta & \alpha - I & \beta & 0 & 0 \\ \multicolumn{6}{c}{\dotfill} \\ \multicolumn{6}{c}{\dotfill} \end{vmatrix} = 0$$

and by substituting $X = (\alpha - I)/\beta$ it is further simplified to

$$\begin{vmatrix} X & 1 & 0 & 0 & 0 & X \\ 1 & X & 1 & 0 & 0 & 0 \\ 0 & 1 & X & 1 & 0 & 0 \\ \multicolumn{6}{c}{\dotfill} \\ \multicolumn{6}{c}{\dotfill} \end{vmatrix} = 0$$

an equation of the 6th degree in X having six roots $m_1, \ldots, m_6$. Substitution of the roots m_i in the secular equations yields the coefficients of the six possible orbitals, and one shall observe that the coefficients depend neither on α nor on β. For a respresentation of the ground state of benzene each of the lowest three orbitals must be counted twice, and this approximation of benzene gives the energy

$$2(\varepsilon_1 + \varepsilon_2 + \varepsilon_3),$$

and since

$$\varepsilon_1 = \alpha - m_1\beta, \qquad \varepsilon_2 = \alpha - m_2\beta, \qquad \varepsilon_3 = \alpha - m_3\beta$$

we also find

$$E^0 = 6\alpha + 2(m_1 + m_2 + m_3)\beta.$$

In general, the physical properties of benzene can be expressed as functions of α and β. One has the freedom to choose empirical values for α and β so as to reproduce certain experimental findings. Those are the principal elements of *Hückel's empirical method* [38]. It is applicable to each conjugated, plane molecule. The parameters may be assumed constant over a series of related molecules. Adaptations of the parameters for one particular molecule then enables us to make predictions about the others.

Sandorfy and Daudel [39] proposed an analogous approximation to deal with localized bonding loges. Hoffmann [40] combined the two methods in order to be able to deal with the σ- and the π-orbitals at the same time. His method is usually called the *extended Hückel method*.

It is also possible to introduce some empiricism into those methods in which the Hamiltonian is formulated as in the self-consistent field method or the configuration interaction method. This will lead to the calculation of a variety of integrals such as

$$\int \chi_i^*(M) \frac{e^2}{r_M} \chi_j(M)\, dv_M, \qquad \int \chi_i^*(M) \frac{h^2}{8\pi^2 m} \nabla \chi_j(M)\, dv_M$$

$$\int \chi_i^*(M) \chi_j^*(M') \frac{e^2}{r_{MM'}} \chi_k(M) \chi_1(M')\, dv_M\, dv_{M'}.$$

The method of Pariser, Pople and Parr [41] neglects some of these integrals and estimates some others through a calculation in which π-orbitals only are determined.

The CNDO method [42] generalizes the preceding technique to the case that σ- and π-orbitals are simultaneously determined. It would require a detailed technical description to do justice to these methods. As such is not the purpose of the present text, the reader is referred to references [19, 43a–b].

CHAPTER III

APPLICATIONS

We analyzed the fundamental ideas at the basis of quantumchemistry, we gave a general survey of the techniques which permit us to effectively calculate wave functions and which at the same time gave us the tools to compose a quantitative language on the basis of the structural elements supplied in the first part of this book.

We have yet to give examples of results arrived at through the application of those methods.

We are going to examine first of all the contributions by that language to an understanding of the structure of de-localized bonds. The chemical notation is not particularly well-adapted to the representation of such bonds. The contributions of quantumchemistry to the subject have been extremely valuable. We will now present some aspects of the quantum theory of the chemical reactivity, while continuing to emphasize the relationships with the nature of the chemical bonds involved. From a quantitative point of view such a study must deal with equilibrium constants and reaction rates. We will demonstrate that quantum chemistry contributes to the solution of important theoretical and practical problems by allowing us to make a fair estimate of those quantities. We will touch its impact on our understanding of the origins of life on our planet, and of the evolution of vegetal and animal life through the ages. We will see how quantumchemistry proves it value, day after day, in pharmacology. We will indicate how purely theoretical processes permit the prediction of novel chemical reactions later to be experimentally confirmed, and eventually we will give an example of an application of quantumchemistry to the aetiology of cancer.

12. The Method of the Molecular Diagram and the Structure of De-localized Bonds

STATIC INDICES

A rather detailed presentation of the Hückel method was given in Chapter II. Since it is our intention that the present chapter can be read without first reading Chapter II, we will now summarize the 'philosophy' behind the method. Consider a benzene molecule. According to Figure 23, Section 6, this molecule comprises one loge with six de-localized electrons and extending over the carbon cores. As this loge is located in the outer regions of the molecule, one may believe that it determines the molecule's reactivity to a great extent. In order to predict this reactivity in a simple way we will use a model in which cores and localized bonds have been replaced by a continuous charge distribution, and with the six electrons kept in the electrostatic field generated by this charge distribution. The mutual interactions of the electrons are neglected in accordance with the independent-electron model (Section 9) and the de-localized bond will be represented by a function derived from a product of one-center functions, called molecular orbitals, obeying the relation

$$(T + F)\, \varphi = \varepsilon \varphi, \tag{3.1}$$

where T denotes the kinetic-energy operator of one electron and F the potential energy it experiences from the charge distribution. It can be shown that within these crude though useful approximations the bond energy E^0 (of benzene, in the present case; the energy of cores and localized electrons has been chosen as the zero on the energy scale) simply equals the sum of the energies associated with the orbitals used for the description of the wave function. It can also be shown that each orbital can be used twice only. For the representation of the ground state of benzene, then, each of the functions φ_1, φ_2, φ_3, assuming that ε_1, ε_2, and ε_3 are the lowest eigenvalues of Equation (3.1), must be used twice. One should bear in mind that degeneracies remain possible and have to be taken into account whenever necessary. Suppose, for instance, that the second eigenvalue is degenerate; then it suffices to note that $\varepsilon_2 = \varepsilon_3$ which φ_2 and φ_3 remain distinct, in order to be able to maintain the entire formalism we are developing. The function associated with the delocalized bond in benzene thus led to the product

$$\varphi_1(M_1)\, \varphi_1(M_2)\, \varphi_2(M_3)\, \varphi_2(M_4)\, \varphi_3(M_5)\, \varphi_3(M_6)$$

and the energy

$$E^0 = 2\varepsilon_1 + 2\varepsilon_2 + 2\varepsilon_3.$$

The difficulties encountered when one attempts at a solution of Equation (3.1),

$$h\varphi = (T + F)\, \varphi = \varepsilon \varphi,$$

imposes the introduction of a new approximation whereby the exact form of the orbital φ is replaced by a linear combination of the atomic orbitals Lp_0 (the z-axis perpendicular to the plane of the hexagon) of each of the carbon cores. With χ_i for the Lp_0 associated with the i-th carbon core we have

$$\varphi_j = c_{1j}\chi_1 + c_{2j}\chi_2 + c_{3j}\chi_3 + c_{4j}\chi_4 + c_{5j}\chi_5 + c_{6j}\chi_6 = \sum_i c_{ij}\chi_i.$$

More or less elaborate methods studied in Section 11, allow us to calculate the coefficients c_{ij} and the energies ε_j.

Suppose now that the orbitals Lp_0 of the various carbon atoms have reasonable values in distinct regions only. Be V_i the volume in which the orbital χ_i is localized. One now proceeds to the calculation of the electronic charge contained therein. By extending the arguments set forth in Section 9, it will not be difficult to find that the probability density of any of the cores and electrons in point M can be written as

$$\varrho_1(M) = \tfrac{1}{3}[\varphi_1^2(M) + \varphi_2^2(M) + \varphi_3^2(M)].$$

As we have a total of six electrons, the total density will be

$$\varrho(M) = 6\varrho_1(M) = 2[\varphi_1^2(M) + \varphi_2^2(M) + \varphi_3^2(M)],$$

and the electronic charge q_1 contained in V_i

$$q_i = \int_{V_i} \varrho_i(M)\,dv$$

can easily be expressed as a function of the c_{ij}. Calculating the contribution to this charge by orbital φ_j one finds

$$\int_{V_i} |\varphi_j|^2\,dv \simeq |c_{ij}|^2 \int_{V_i} |\chi_i|^2\,dv = c_{ij}^2$$

because of the normalization. So eventually

$$q_i = 2|c_{i1}|^2 + 2|c_{i2}|^2 + 2|c_{i3}|^2.$$

For benzene the customary methods give $q_i = 1$. The de-localized bond introduces (equal) unit charges to each of the cores of benzene. It can be shown, though in a less clear manner, that the quantity

$$p_{k1} = 2[2c_{k1}c_{l1} + 2c_{k2}c_{l2} + 2c_{k3}c_{l3}],$$

the so-called *bond order* [44a–b], is proportional to the energy contained in the de-localized loges' fraction that extends between the two adjacent cores k and l.

MOLECULAR DIAGRAMS OF CONJUGATED HYDROCARBONS

The conjugated hydrocarbons form a group in which the static indices defined in the preceding section prove extremely useful. The purely conjugated (i.e. non-substituted) hydrocarbons are molecules formed from carbon and hydrogen and can be represented by formulae in which single and double bonds between the carbon atoms alternate. There are the polyenes (butadiene,

butadiene naphtalene fulvene

hexatriene etc.), the acenes (benzene, naphtalene, phenantrene) and a good many other compounds such as fulvene and azulene. A good division into loges like in the case of benzene, requires a de-localized n-electron bonding loge (for a molecule with n carbon atoms) around the carbon cores and the localized single bonds. This is indicated in Figure 32, whereas the commonly used chemical symbols are given in Figure 33. Figure 34 shows the charge distribution (calculated with the Hückel method, Chapter II) arising from the de-localized bonds, in the vicinity of the carbon cores. Notice that for the first two molecules these charges are unit charges everywhere. In the third molecule, however, the extreme atom is less rich in electrons than the ones in the ring. This leads to a polarity in the molecule: positive at the extreme atom, negative in the ring. An electric dipole is bound to appear. This has been indicated in

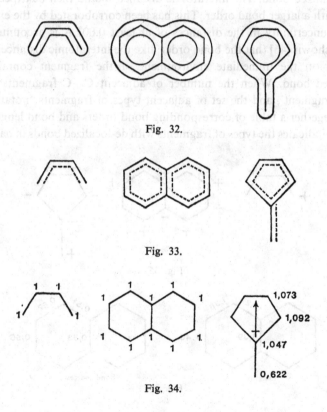

Fig. 32.

Fig. 33.

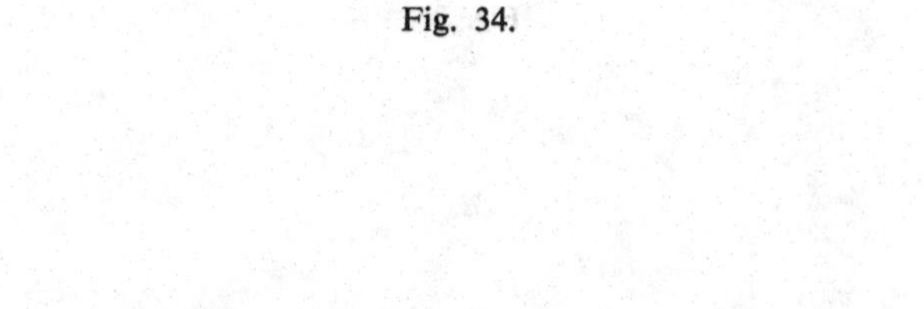

Fig. 34.

Figure 34. Experiments confirmed these results. Butadiene and naphtalene have practically no dipole moment, but fulvene has one of 1.2 D. An examination of the topology of the de-localized bonds also permits us to distinguish between two types of conjugated hydrocarbons, and to predict without a calculation the structure of those bound to have a dipole. It suffices to label the carbon cores by + or − signs (interpreted as indicating $+\frac{1}{2}h/2\pi$ or $-\frac{1}{2}h/2\pi$ spin) in such a way that two adjacent signs shall not be identical, whenever possible. In case this can be carried out with success the hydrocarbon is *alternant*, if not it is *non-alternant*. Figure 35 shows that butadiene and naphtalene are alternant whereas fulvene is non-alternant. It is easy to see that the presence of rings with an odd number of carbon atoms prevents a hydrocarbon from being alternant. It can be shown that the charges in all alternant hydrocarbons are unit charges [45]. As a result of that they will have very small dipole moments. In non-alternant hydrocarbons the charges usually are different from unity. The dipole moments will therefore be considerable. In Figure 36 the lengths of the C—C bonds can be compared with their bond orders. We noted before that the bond order is a kind of yardstick for the energy of the fragment corresponding with the de-localized bond. The interatomic distance should then be expected to be shorter with a larger bond order. This has been corroborated by the experiment when an uncertainty in the distances of at least 0.007 Å is accounted for. It has been shown [46] that the bond order, like the interatomic distance, depends above all on the immediate environment of the fragment containing the de-localized bond. When the number of adjacent C—C fragments is called 'type of fragment', and the set of adjacent types of fragments 'notation', one can put together a table of corresponding bond orders and bond lengths [43a]. Figure 37 indicates the types of fragments with de-localized bonds in naphtalene.

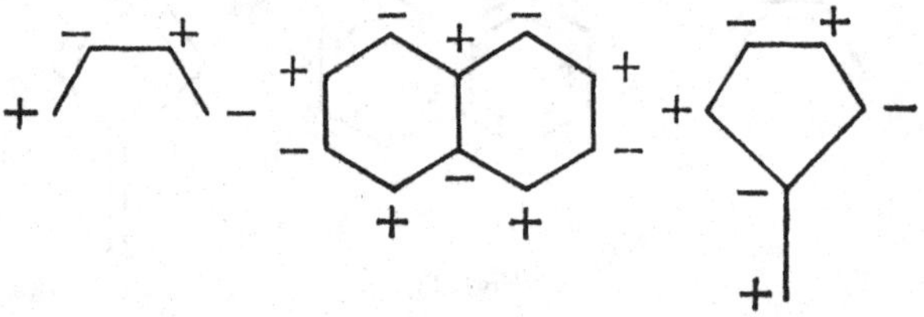

Fig. 35.

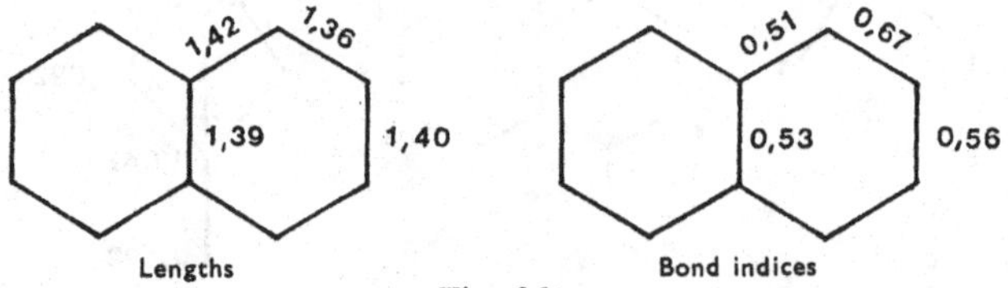

Lengths   Bond indices

Fig. 36.

It explains why the fragment $\alpha\beta$ has the notation (2, 3) and the central fragment the notation (3, 3, 3, 3).

We observe, for instance, that a (2, 3)- or a (3, 3) bond is short: 1.37 ± 0.02 Å; that a (3, 3, 4) bond is medium: 1.40 ± 0.01 Å; that a (4, 4, 4, 4) bond is long: 1.44 ± 0.02 Å.

It is convenient to add some more static indices to the ones introduced so far. We will begin with the free-valence index [47a–c], which quantity adds the flavour of mathematics to the concept of residual affinity which has been on the chemists' minds for such a long time. In order to arrive at this concept one has to assume that a given carbon atom can be held responsible for only a limited part of the bonding energy. The sum of the bond orders of the delocalized fragments emanating from this atom, may be considered as the measure to which the atoms contributes to the total energy. The difference between a certain constant and the sum of the bond orders

$$F_1 = C - \sum_m p_{1m}$$

measures the residual affinity of the atom and constitutes its free-valence index. The constant C is arbitrary. Very often it is given the value 1.732. In Figure 38 the values for naphtalene are given.

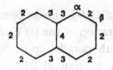

Fig. 37.

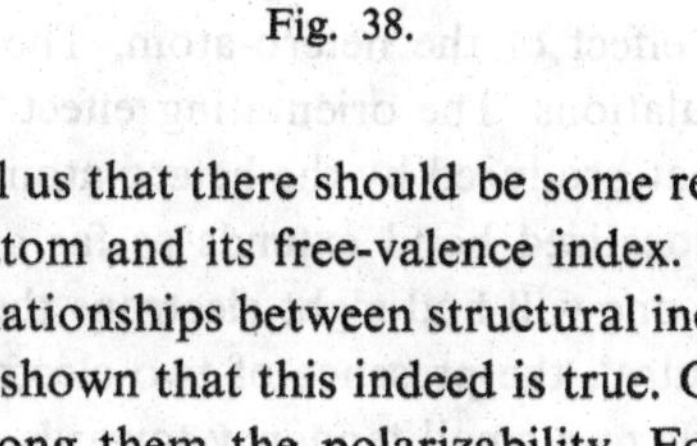

Fig. 38.

Our intuition may tell us that there should be some relationship between the reactivity of a carbon atom and its free-valence index. When we lay the basis for the mathematical relationships between structural indices and reaction rates in Section 13, it will be shown that this indeed is true. Other indices have been introduced as well, among them the polarizability. For them, and for more details about the indices analyzed here, the reader is referred to more technical texts [19, 43a]. The only thing we like to add here is that a chemical formula with the static indices inserted, either in numbers of in a symbolic notation, is usually called a *molecular diagram* [48].

MOLECULAR DIAGRAMS OF OTHER CONJUGATED MOLECULES; ORIENTATING EFFECT OF SUBSTITUENTS AND HETERO-ATOMS

Numerous molecules can be considered as derivatives of purely conjugated hydrocarbons, either through replacement of a CH-group by a hetero-atom or by substitution. Figure 39 indicates how pyridine can be thought of as being the

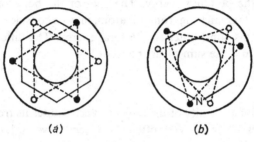

(a)   (b)

Fig. 39.

result of the replacement of a CH-group in benzene by nitrogen. Figure 39a gives a very probable configuration of the six electrons in the de-localized loge of benzene, the electrons with parallel spins being at the largest possible distance in the corners of an equilateral triangle. The similar situation prevailing in pyridine is given in Figure 39b. In this case the electrons with parallel spins occupy the corners of such a triangle again. Within the de-localized loge, however, a nitrogen atom is found with 2 electrons most of the time in the vicinity of the nucleus with a charge $+7e$. Its apparent charge is then $+5e$, whereas the carbon cores usually carry a charge of only $+4e$. The electrons will therefore have a tendency to go closer to the nitrogen, its charge q_N exceeding 1.

The quasi-triangular correlation between the electrons with parallel spins leads to a correlated decrease of the charge at the ortho- and para-positions. This is the orientating effect of the hetero-atom. The diagrams in Figure 40 give the results of calculations. The orientating effect of a OH-substituent is evidently opposite to that produced by the hetero-atom. This is due to the fact that in phenol the de-localized bond extends as far as the oxygen core. The probability is high that one will find eight electrons there. For the region next to the oxygen to be neutral, the presence of two electrons would be required. The de-localization leads to a small tendency towards a more equitable distribution of the electrons in the loge. The charge q_0 goes down to 1.936 (instead of

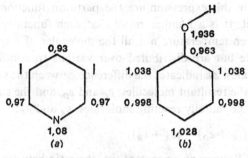

Fig. 40.

2) and the resulting small transfer of electrons to the ring concentrates in the first place on the ortho- and para-positions.

GRAPHITE AND THE METALLIC BOND

Figure 41a shows how the enormously large de-localized bonding loge in graphite extends over all the carbon cores, and branches out to form a multitude of paths suitable for a collective movement of the electrons under the influence of a potential differential at the extreme ends of the molecule. This is the explanation of the electrical conductivity of graphite. It can be done likewise for metals. In Figure 4lb is shown the de-localized bond in metallic lithium, extending in three dimensions over the K-loges of each nucleus, without any other limitation than an edge of the crystal or a stacking fault.

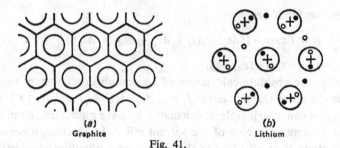

(a)
Graphite

(b)
Lithium

Fig. 41.

13. Chemical Equilibria; Biological and Pharmacological Applications

Let

$$A + B \overset{K}{\rightleftharpoons} C + D$$

be a reversible reaction. In so far as concentration and activity may be confounded, the equilibrium constant K will be given by the expression [30]

$$K = (f_C f_D / f_A f_B) \exp - (\Delta \varepsilon / kT) .$$

The functions f in this expression are the partition functions of the various molecules present. It is a familiar result that such functions account for the fact that at a given temperature not all the molecules of a species are in the same energy state but are distributed over various translational, vibrational and librational states. $\Delta\varepsilon$ indicates the difference between the sum of the ground state energies of the resultant molecules, ε_C and ε_D, and the sum of the ground state energies of the initially reacting molecules, ε_A and ε_B:

$$\Delta\varepsilon = (\varepsilon_C + \varepsilon_D) - (\varepsilon_A + \varepsilon_B).$$

The quantity k is Boltzmann's constant (i.e. the ratio between the gas constant of an ideal gas and Avogadro's number). The usual methods separate the energy of a molecular state into five terms:

(a) ε_v = the minimum nuclear vibration energy;

(b) ε_1 = the energy of the cores and the localized bonds;

(c) ε_d = the energy of the de-localized bonds in the field of the cores and the energy of the localized bonds;

(d) ε_{n1} = the interaction energy between non-bonded atoms, usually neglected for the calculation of the preceding terms, and

(e) $\varepsilon_s(T)$ = the solvation energy.

This latter term, accounting for the solvent effect, is the only one depending on the temperature. Even when in its ground state, the molecule interacts with the solvent in a way essentially depending on the thermal agitation. Writing

$$f = (f_C f_D / f_A f_B)$$

the expression for K becomes

$$K = f \exp - ((\Delta\varepsilon_v + \Delta\varepsilon_d + \Delta\varepsilon_1 + \Delta\varepsilon_{n1} + \Delta\varepsilon_s(T))/kT),$$

where $\Delta\varepsilon_i = (\varepsilon_{Ci} + \varepsilon_{Di}) - (\varepsilon_{Ai} + \varepsilon_{Bi})$.

Evidently, the absolute calculation of an equilibrium constant requires the exact calculation of the six factors f, $\Delta\varepsilon_v$, $\Delta\varepsilon_d$, $\Delta\varepsilon_1$, $\Delta\varepsilon_{n1}$ and $\Delta\varepsilon_s(T)$. Each one of these terms can in principle be calculated by wave mechanics, combined with statistical mechanics in case of the solvent effect. In practice, however, it can hardly be done if at all. This is the reason why quantum chemistry is well equipped to compare rate constants, but less so to accurately compute them. Look at two analogous reactions between similar molecules

$$A + B \overset{K}{\rightleftarrows} C + D \quad \text{and} \quad A' + B' \overset{K'}{\rightleftarrows} C' + D'.$$

In an obvious notation we have

$$K'/K' = (f'/f) \exp - (\Delta\Delta\varepsilon_v + \Delta\Delta\varepsilon_d + \Delta\Delta\varepsilon_1 + \Delta\Delta\varepsilon_{n1} + \Delta\Delta\varepsilon_s(T))/kT.$$

It often so happens that in a family of related molecules one or two $\Delta\varepsilon$ change much more going from one molecule to another, than any of the other terms. In such cases the expression for K will be dominated by one or two $\Delta\Delta\varepsilon$. In essence they will determine the variation of the relative constants K'/K throughout the family.

THE ALKALINITY OF CONJUGATED MOLECULES IN THEIR ELECTRONIC GROUND STATE

The study of the variation of the pK in the family of amino-derivatives of pyridine, isoquinoline, quinoline and acridine will serve as an example. For pyridine the equilibrium constant K_a corresponds with the reaction

$$\left[\underset{\underset{H}{N}}{\bigcirc}\right]^+ \underset{K_a}{\rightleftharpoons} \underset{N}{\bigcirc} + H^+$$

The negative logarithm of K_a is pK. A higher value of pK means that a molecule is more alkaline, i.e. it is more readily inclined to accept a proton. When we pass from pyridine to acridine we will have to consider the reaction

$$\left[\underset{\underset{H}{N}}{\bigcirc\bigcirc\bigcirc}\right] \underset{K'_a}{\rightleftharpoons} \underset{N}{\bigcirc\bigcirc\bigcirc} + H^+$$

It is easy to see that among the $\Delta\Delta\varepsilon$ the term $\Delta\Delta\varepsilon_d$ may be very important as the de-localized bond in acridine differs a great deal from the one in pyridine; it covers three rings. For that reason it is quite possible that the fixation of a proton to the nitrogen will modify to an entirely different extent the energy of the de-localized bonds in pyridine and acridine. A convincing argument could be the equality holding true for the simple methods,

$$\Delta\varepsilon_d = q_N \Delta\alpha,$$

where q_N is the charge carried by the de-localized bond to the vicinity of the nitrogen core, and $\Delta\alpha$ the change in the Coulomb integral of the nitrogen caused by the fixation of the proton. $\Delta\varepsilon_d$ must then vary with the charge on the nitrogen. Then the pK may be expected to vary with that charge, in perfect agreement with chemical intuition. In fact, one may expect pK_a, the alkalinity, in that family of molecules to increase with the charge on the nitrogen, since the greater the latter is, the greater will be the nitrogen's tendency to accept a proton. A similar relationship is found between pK and q_N in a family of derivatives with not too different sizes. Compare, as an example, pyridine and para-amino pyridine,

$$NH_2$$

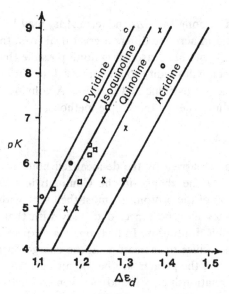

Since the NH_2-group behaves like the OH-group, it must be expected (see Section 12) that in the latter compound the charge q_N shall be greater than in pyridine. This was confirmed by a calculation. The pK in an aqueous solution at 20 °C is 5.23 for pyridine and 9.17 for para-amino pyridine. A comparison of derivatives with widely different sizes will prove to be far more complicated. In such cases it is preferable that $\Delta \varepsilon_d$ is calculated using a rather elaborate method such as the Pariser-Parr-Pople method. The variation of pK as a function of $\Delta \varepsilon_d$ is given in Figure 42 [49]. The existence of a good relationship between pK and $\Delta \varepsilon_d$ (and so between pK and the charge q_N) for each group of amino-derivatives of the same molecule, is clear. The relationship changes, however, going from one base to another. This observation suggests that at least one more $\Delta \Delta \varepsilon$ must be important; knowing that the solvation energy often depends on the molecular dimensions, one is led to calculate $\Delta \Delta \varepsilon_s(T)$. The complete calculation of this term is by no means an easy task. However, the analysis of the principal factors determining the solvation energy, indicates that in the cases we are interested in, the electrostatic interaction between the

Fig. 42.

charges of the de-localized bond and the solvent molecules plays the leading role. For a molecule containing the charges q_i an estimate can be made with the formula [50]

$$\varepsilon_s(T) = -\tfrac{1}{2} \sum_i (q_i - \eta)^2 \, r_i^{-1} (1 - D^{-1}).$$

In this expression $(q_i - \eta)$ represents the apparent charge of an atom. Thus $\eta = 2$ for the nitrogen in the NH_2-group and $\eta = 1$ for a heteroatomic nitrogen and for carbon atoms. r_i is the apparent radius of the atom in the solvent (i.e. the distance to which the solvent molecules can approach the atom) and D the effective dielectric constant. It is shown in Figure 43 that for the whole

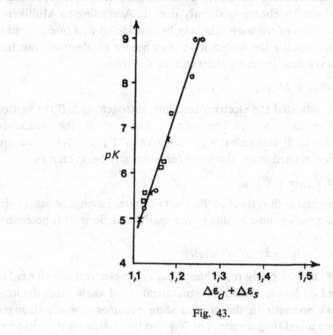

Fig. 43.

family of molecules studied, one single straight line is obtained when pK is plotted as a function of $\Delta\varepsilon_d + \Delta\varepsilon_s(T)$. A more refined study of the problem will reveal that $\Delta\varepsilon_{n1}$ has some significance too.

THE ALKALINITY OF CONJUGATED MOLECULES IN THE FIRST EXCITED ELECTRONIC STATE

In Table V have been brought together a number of measured pK's in the ground state, first excited singlet state and first triplet state. A big change of pK is observed going from the ground state to the first excited singlet state. Depending on the molecule, it may either decrease or increase. The pK's of the triplet state, on the other hand, are only slightly different from those pertaining

TABLE V

Molecule	Ground state	First excited singlet	First triplet
β-naphtol	9.46	2.8	8.1
β-naphthylamine	4.1	−2	3.3
Acridine	5.5	10.6	5.6

to the ground state. Quantummechanically, it is not difficult to understand the quantitative origin of this phenomenon [51a–c]. Take acridine for example. We know that in this molecule the nitrogen attracts the electrons more strongly than the carbon does; its charge is slightly over 1. According to Mulliken's ideas, the wave function of the molecule may be considered as a linear combination of Ψ_{AD}, representing the molecule in the absence of electron transfer, and $\Psi_{A^-D^+}$, the wave function with electron transfer. Then

$$\Psi = a\Psi_{AD} + b\Psi_{A^-D^+} .$$

In this notation A indicated the electron acceptor, nitrogen, and D the rest of the molecule, acting as a donor. The transfer being small, the inequality $|a| \gg |b|$ must hold, i.e. Ψ resembles Ψ_{AD} rather than $\Psi_{A^-D^+}$. When we are interested in the first excited state, the wave function may be written as

$$\Psi' = a'\Psi_{AD} + b'\Psi_{A^-D^+}$$

and must be orthogonal with respect to Ψ. The two states having the same spin function, the orthogonality must be due to the spatial functions. It is necessary that

$$aa' + bb' = 0, \quad \text{or} \quad |b'/a'| = |a/b|.$$

This means that Ψ' now strongly resembles $\Psi_{A^-D^+}$, the electron transfer is far more extensive and q_N has increased. A generalization of these ideas demonstrates that a weak acceptor in the ground state becomes a much stronger acceptor in the first excited singlet state. The fact that the charge on the nitrogen is greater in the first singlet excited state than in the ground state, allows us to understand why the pK is larger in the excited state. Turning around the argument, it can be shown that a weak donor in the ground state (like NH_2 and OH) becomes a stronger donor in the first excited singlet state.

It can now also be understood why excitation of β-naphtylamine and β-naphtol, contrary to acridine, leads to a lower pK when the spin remains unchanged. It still has to be understood why the triplet and the first excited singlet states have such very different pK's. Modern computational methods lead to very similar charges for those lower states. It seems that the origin of the difference should be found elsewhere. We know that an approaching proton polarizes a molecule, i.e. modifies its charge distribution (see Section

8). The problem may be simplified by the discussion of a two-electron system. The singlet state is characterized by a symmetric space function, $\Psi^S(M_1, M_2)$, and the triplet by an anti-symmetric function $\Psi^A(M_1, M_2)$. The probability of finding two neighbouring electrons in the triplet state will be very small. as

$$\Psi^A(M_1, M_2) = -\Psi^A(M_2, M_1)$$

and therefore

$$\Psi^A(M, M) = -\Psi^A(M, M) = 0.$$

In this case the proton is unable to promote the presence of two electrons in the de-localized loge in the vicinity of the same core. The molecule has a low polarizability. This argument does not hold for the singlet. The conclusion has to be that the triplet states can be expected to have a lower polarizability than the singlet states (ground or excited). The result is that the nitrogen atoms of two acridine molecules, one in the triplet state and the other in the first excited singlet state, will not undergo the same changes under the influence of an approaching proton, even when they initially have the same charge q_N. The nitrogen in the singlet state has a higher polarizability and will undergo a far more considerable charge increase. Its pK will be larger.

THE ORIGIN OF LIFE ON EARTH; EVOLUTION OF THE SPECIES

According to current ideas, living nature distinguishes itself from inanimate matter by its aptitude to develop, reproduce and subsequently die. The impression is given that a living creature has been organized so as to accomplish a task. Jacques Monod[52] wrote: 'We shall arbitrarily define the teleonomical project as the transmission from one generation to another of the amount of invariance characterizing the species', and Jacob speaks of the 'dream' each cell has, of becoming two others, sometime. It is interesting to note that biologists often indulge in a flirtation with teleology, whereas physicists adopt a more positive attitude by nearly always avoiding such finalistic thinking. The epistemology of physics creates an entirely different atmosphere from the one created by the epistemology of biology. The essential difference should bring us closer to an understanding of what life really is. Among the questions raised in that connection, the ones concerned with the origin of life on our planet and its diversification, undoubtedly take a place on the forefront. We like to note that quantumchemistry made a very modest contribution to the subject by inciting young workers to research into that direction. At present we know that the genetic code allowing of the evolution of a living being and the transmission of information as mentioned by Monod, is contained in the nucleic acids. These acids contain a chain of four organic bases in a certain order: adenine, guanine, thymine (or uracil) and cytosine. Desoxyribonucleic acid, or DNA, contains two parallel and helicoidal chains of purine- and pyrimidine bases. Always, a thymine stands opposite an adenine, a cytosine opposite a guanine, and the pairing is ensured by hydrogen bonds. The structure of DNA is described by Figure 44. For the sake of simplicity the various bases have been indicated by *A, C, T* and *G*. A DNA-fragment may be symbolized for example, by the notation

$$A\ T\ T\ C\ G\ A\ C\ T$$
$$T\ A\ A\ G\ C\ T\ G\ A$$

and the genetic code is embedded in the order of the bases in the chains. One of DNA's fundamental properties is its ability to reproduce in a suitable environment, but particularly in the presence of an enzyme called polymerase. The two chains of the double helix unfold and the bases place themselves in front of the free fragments, always respecting the pairing of *A* with *T* and of *C* with *G*. A schematic picture of this operation is given in Figure 45. The DNA born out of this very simple procedure contains exactly the same 'message' as the initial DNA; this is *replication*. It is possible to understand in this way how in the course of cell division one cell produces two cells carrying the same message as the original one. This is the basis of the scientific explanation of the reproduction of the majority of living beings. A second fundamental property of DNA is its ability to have only a part of itself, a gene, transcribed in the presence of the enzyme DNA-dependent RNA polymerase. In this manner one molecule of ribonucleic acid, RNA, is formed. Generally comprising one

Sugar—Adenine ... Thymine—Sugar

Phosphate   Phosphate

Sugar—Thymine ... Adenine —Sugar

Phosphate   Phosphate

Sugar—Cytosine ... Guanine—Sugar

Fig. 44.

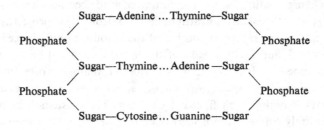

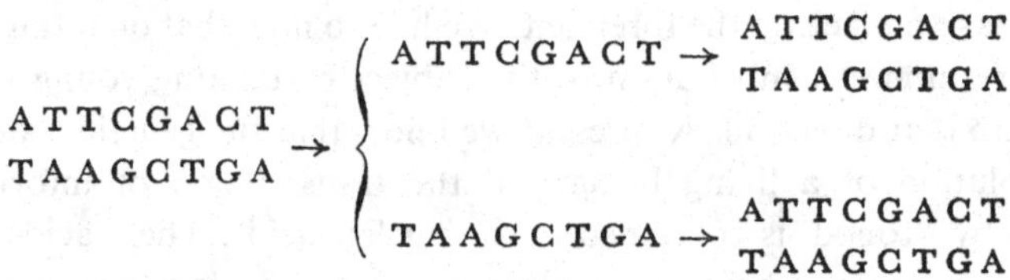

Fig. 45.

chain only, this molecule also distinguishes itself from DNA by containing uracil instead of thymine. Suppose now that the fragment *ATTCG* of the abovementioned DNA element, constitutes such a gene; then it will be capable of inducing the formation of a RNA molecule according to the following scheme:

$$\text{fraction of DNA chain} = A\,T\,T\,C\,G$$
$$\text{RNA} = U\,A\,A\,G\,C.$$

This is the *transcription*. The various RNA molecules thus formed are able to incite the cell machinery to manufacture specific molecules. Different genes of DNA can engender different types of RNA, each one of the latter being capable of inciting the cell to manufacture one particular kind of molecule. This may have given the reader a feeling of how the genetic code will be translated into a sequence of syntheses, and how step by step an individual and its reproduction will be brought about.

In order to understand how life came to Earth, one has to have an understanding of the formation of the first nucleic acids and the first enzymes. Nucleic acids contain purine- and pyrimidine bases, enzymes contain amino acids; it shall be our first task then to investigate how such molecules originated. For a scientific attempt at the reproduction in vitro of the events on Earth in the remote past to be significant, we must first formulate an opinion on the nature of the substances present at that time.

Contemporary cosmology [53] tends to assume that the Universe is the result of approximately twelve billion years of evolution of a hydrogen cloud. Our galaxy would then be about seven billion years old and the solar system about five billion years. Rather precise studies mark the appearance of bacteria at three billion years ago. It seems probable that at those times the terrestrial atmosphere was a good deal different from the present one. We have important reasons to believe that its predominant constituents were hydrogen, methane, ammonia and nitrogen. The methane was gradually replaced by carbondioxide, then the oxygen appeared probably three hundred million years ago, undoubtedly as a result of photosynthesis. In 1953 Miller [54] subjected a mixture of methane, ammonia, water vapour and hydrogen to a substantial electrical discharge and obtained glycine and alanine, i.e. two amino acids. The constituents of the proteins can apparently be formed under very simple conditions from the mineral substances probably present on Earth at the time that life originated. The same molecules may be obtained by irradiation with ionising radiation of a similar mixture. Ponnamperuma *et al.* [55] obtained adenine in 1963 by irradiation of a mixture of methane, ammonia and water vapour. Adenine and a small amount of guanine was obtained later by ultraviolet irradiation of a solution of hydrocyanic acid. How is it possible that the irradiation of molecules as simple as CH_4, NH_3, H_2O and HCN gives rise to molecules as complicated as adenine and guanine, with the formulae

At this very stage, quantum chemistry contributes its first elementary answer [56a, b]. It is actually possible to calculate the binding energy of various arrangements of atoms of carbon, nitrogen, hydrogen and oxygen. It was found that the purine- and pyrimidine bases belong to the most stable configurations, and that the order of the stability is

adenine > guanine > cytosine > thymine.

(From a more thorough analysis of this problem we would have learned that the presence of a long de-localized bond contributes substantially to this stability.) In the experiments mentioned above, the very formation of adenine, and sometimes of guanine, was found. It is now possible to formulate a first idea about the phenomenon by supposing that the radiation degrades the molecules, and that subsequently the fragments reorganize into more stable assemblies. But once it has been assumed that a living individual, bacteria for instance, has been established it will not suffice to know how it is capable of reproduction, of transmission of the invariant characteristics of its species from one generation to the next, should we wish to understand the origin of all the vegetal and animal species present in our biosphere. We have to come to an understanding of the way in which an individual of a given species may engender an individual of a different species as the result of an 'error' (or 'predestination', dependent on the metaphysical framework one prefers). We know that the phenomenon of mutation provides the basis for the scientific interpretation of such an event. A veritable mutation is a fundamental error in the formation of a conjugated chain. The newly formed chain will then carry a genetic code which differs from the original message. A new species may be the result. Here again, quantumchemistry suggests possible roads leading towards the error. Consider the normal pairing of cytosine and guanine as given in Figure 46.

Suppose that radiation brings the cytosine to its first excited state. We know that the hetero-atomic nitrogen will become a stronger acceptor. It will attract protons with more vigour. The nitrogen in the amino group, on the other hand, will become a stronger donor, more positive. It tends to loosing a proton. Some of the protons in the amino group have a tendency to jump to the hetero-atom

nitrogen

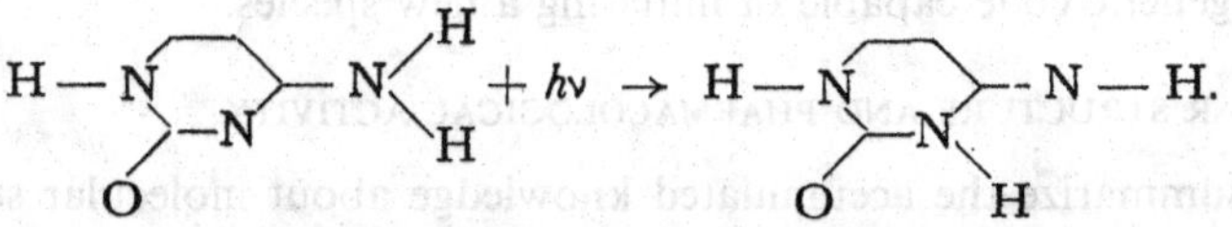

$$H-N \underset{O}{\overset{}{\bigcirc}} -N \overset{H}{\underset{H}{\diagdown}} + h\nu \rightarrow H-N \underset{O}{\overset{}{\bigcirc}} -N-H.$$

The resulting molecule is badly equipped for a linkage to the guanine, although it is still capable of pairing with an adenine molecule. This is shown in Figure 47. This line of reasoning follows almost exactly the ideas put forward by A. Pullman [57], and the conclusion to be drawn is that replication of DNA in the presence of radiation, creates a chance for cytosine to place itself in front of an adenine molecule. The scheme of Figure 45 may well become that of Figure 48. Notice that the double chain in the top right-hand side of this

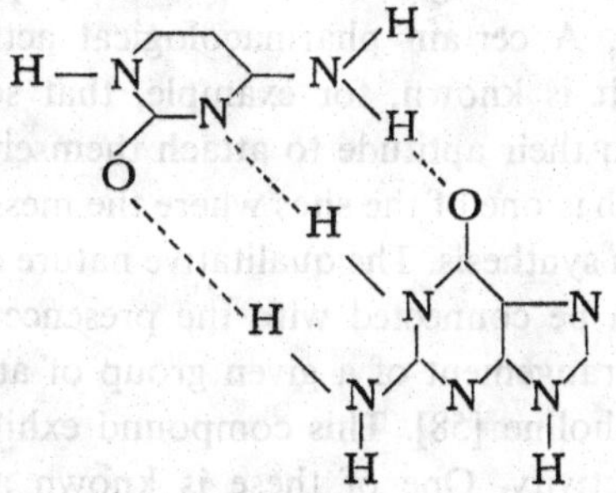

Fig. 46.

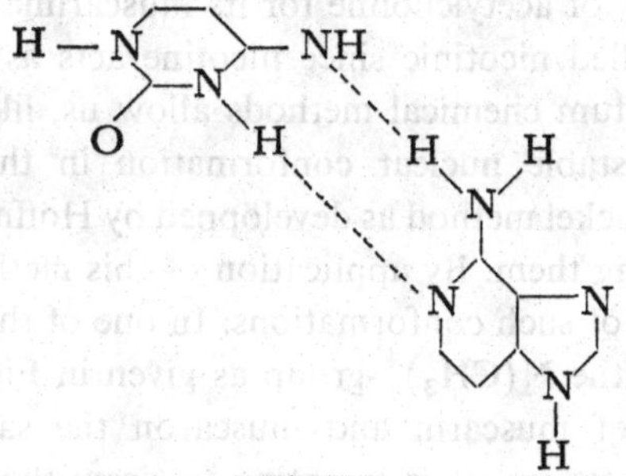

Fig. 47.

ATTC*GACT → ATTCGACT
            TAAACTGA

ATTCGACT →
TAAGCTGA

TAAGCTGA → ATTCGACT
            TAAGCTGA

Fig. 48.

figure contains a sequence of three A's instead of two. It contains an error, a different genetic code capable of initiating a new species.

MOLECULAR STRUCTURE AND PHARMACOLOGICAL ACTIVITY

We can summarize the accumulated knowledge about molecular structure by saying that a molecule is the result of a kind of self-consistency of the spatial arrangements of the nuclei and of the electrons. The nuclear conformation depends on the arrangement of the electrons, but the latter in turn adjust themselves to the field of the nuclei and the interelectronic interaction. The molecular properties may therefore be thought of as to be simultaneously related to the spatial arrangement of the nuclei and of the electrons.

An examination of the relations between structure and pharmacological activity provides evidence of this dualism. A molecule's aptitude to attach itself to a particular site in a given cell, seems to depend first of all on the nuclear conformation. A certain pharmacological action often begins with such an attachment. It is known, for example, that some antibiotics derive their effectiveness from their aptitude to attach themselves to a particular site of the ribosomes, which is one of the sites where the messenger-ribonucleic acid goes during the protein synthesis. The qualitative nature of the pharmacological effectiveness will often be connected with the presence in the molecule of a specific geometrical arrangement of a given group of atoms. Kier reported in great detail on acetylcholine [58]. This compound exhibits two distinct kinds of pharmacological activity. One of these is known as muscarinic activity, since muscaron and muscarin have the same activity. These latter substances are called the agonists of acetylcholine for its muscarinic activity. The second kind of activity is called nicotinic since nicotine acts as the agonist for this activity. Several quantum chemical methods allow us, like we saw before, to determine the most stable nuclear conformation in the electronic ground state. The extended Hückel method as developped by Hoffmann, is undoubtedly the simplest one among them. By application of this method, Kier found that acetylcholine has two of such conformations. In one of these, the two oxygens have the distances to the $N(CH_3)_3^+$-group as given in Figure 49. In the most stable configuration of muscarin and muscaron the same atoms occur at precisely the same distances. It is tempting to draw the conclusion that the geometric arrangement of the atoms in Figure 49, constitutes the muscarinic pharmacophore, i.e. has the spatial arrangement of the nuclei necessary to let the muscarinic activity appear in that kind of molecule. A possible explanation could be that the activity has as its first step the fixation to a site in the cell with a complementary conformation, or at least capable of adopting such a conformation at the moment of formation of the addition complex. The importance of this hypothesis is emphasized by the fact that it has a counterpart in the case

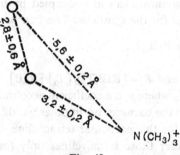

Fig. 49.

of the nicotinic activity. The second stablest conformation of acetylcholine and the stable conformations of nicotine contain the same structural element in the spatial arrangement of the nuclei. However, considering the vastness of the group of molecules containing the same pharmacophore, one wonders which factors are responsible for the quantitative aspects of their pharmacological activity. Quite naturally, one will think in terms of the distribution of the electron density. We will examine tetracycline in connection with the beautiful work of Peradejordi and Martin [59]. The basic formula of this group of molecules is

The antibiotic power of this group is known to be heavily dependent on the nature of the constituents R_1, R_2, R_3, R_4 and R_5. This antibiotic power is measured by comparing the rates of growth of a bacterial cell culture (*Escherichia Coli*). The number of cells changes with time according to the expression

$$N_t = N_0 \exp k_i t \quad \text{where} \quad k_i = k^0 - A_i[T_i].$$

In this latter formula $[T_i]$ represents the tetracycline concentration added to the substrate. The constant k^0 in the exponent represents the cell growth in the absence of antibiotic, and A_i measures the antibiotic activity of the particular tetracycline indicated by the subscript i. Many experimental data suggest that the tetracycline inhibits the protein synthesis, and consequently the synthesis of nucleic acids, by attaching itself to those sites on the ribosomes which under normal circumstances are occupied by transfer-RNA supplying the aminoacids required for the synthesis. This can be written

$$T_i + \text{RIB} \underset{K_i}{\rightleftarrows} \text{RIB}, T_i,$$

where RIB = ribosome and $K_i = [\text{RIB},\text{T}_i]/[T_i]\,[\text{RIB}]$.

We may put $A_i = p_i K_i$, where p_i is a coefficient accounting for the tetracycline's ability to penetrate into the bacteria, or, in other words, depends on the permeability of the cell wall for this particular tetracycline. In order to simplify the problem we will consider those tetracyclines only for which we have good reasons to believe, on the basis of experiments performed with model membranes, that their permeabilities have comparable magnitudes. Knowledge about the variation within the group of the antibiotic activity A_i will under such circumstances be reduced to knowledge about K_i. We know that

$$K_i = f_i \exp - (\Delta\varepsilon_i/kT).$$

With the additional assumption that the f_i vary only slightly (and the results will permit to see whether this is a justifiable assumption) a study of K_i is reduced to a study of $\Delta\varepsilon_i$. This $\Delta\varepsilon_i$ represents the sum of the interaction energies prevailing among the atoms in the tetracycline and those in the ribosome. Consider the pair p of interacting atoms. The corresponding energy may be written

$$\Delta\varepsilon_p = f_p Q_p + g_p E_p + h_p N_p,$$

where Q_p, E_p and N_p are respectively the apparent charge of the atom of the tetracycline, the *electrophilic super-de-localizability* and the *nucleophilic super-de-localizability*. (These latter quantities were introduced by Fukui and are a measure of the kinds of polarizabilities arising from an electrophilic- or nucleophilic attack). The other factors, f_p, g_p and h_p are the analogue quantities of the ribosome. Now we have

$$\Delta\varepsilon_i = \sum_p \Delta\varepsilon_p$$

and so

$$LA_i = \text{constant} + \sum_p (f_p Q_p + g_p E_p + h_p N_p)/kT.$$

Quantumchemical methods enable us to calculate Q_p, E_p and N_p. On the other hand, f_p, g_p and h_p are not accessible to a calculation as it is unknown which atoms of the ribosome interact with those of the tetracycline. We are forced to resort to the empirical tool of determining such coefficients that minimize the difference between the squares of the experimental values of the A_i and those computed with the preceding formula. This led to the expression

$$LA_i = 18.4 + 56Q_{O_{10}} + 17E_{O_{10}} + 48Q_{O_{11}} +$$
$$- E_{O_{11}} + 71Q_{O_{12}} + 18E_{O_{12}} + 3Q_{C_6}.$$

Figure 50 shows that the differences between the values calculated with the formula and those experimentally determined, are actually very small. They correspond with a factor 2 over a group of molecules in which the antibiotic activities vary by a factor of over 10^3. The preceding formula has a great practical value, therefore. In particular, it enables us to calculate from first

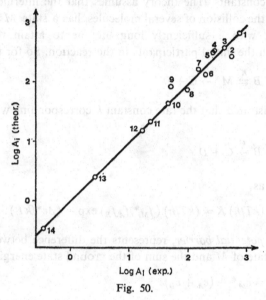

Fig. 50.

principles and with a good precision, the antibiotic activity of a tetracycline after a glance at its formula, and even prior to its actual synthesis. The formula equippes the chemist with some guidance towards the synthesis of a molecule with a desired activity. This method has been tested. Activities of new tetracyclines have been calculated before they were synthesized, and their effectiveness was found in agreement with the theoretical predictions. Bearing in mind that sometimes over a thousand substances have to be prepared in order to find one convenient drug, one will easily recognize the practical importance of the rational methods supplied by quantumchemistry to pharmacology.

14. Reaction Rates; Theoretical Prediction of New Reactions; Application to the Aetiology of Cancer

THE THEORY OF THE TRANSITION STATE

In principle, a chemical reaction is the result of collisions among molecular populations. One could have expected quantumchemistry to rely mainly on the theory of collisions. This is indeed true at the most fundamental level of research. In the field of the relationships between bond arrangements in molecules (i.e. of molecular structure) and the reactivity it entails, the collision theory leads to computational complications that are hard to disentangle. It is preferable to appeal to the theory of the transition state as put forward by Eyring and Polanyi. Their theory reduces the calculation of rates to a calculation of equilibrium constants. The theory assumes that the intermediate complex resulting out of the collision of several molecules, has a state M^*, the so-called *transition state*, with a sufficiently long lifetime to attain thermodynamic equilibrium with the initial participants in the reaction. So for the example of

$$A + B \overset{K}{\rightleftarrows} M^* \rightarrow C + D$$

it can be demonstrated that the rate constant k corresponding with the overall-reaction

$$A + B \overset{k}{\rightarrow} C + D$$

may be written as

$$k = (kT/h) \, K = (kT/h) \, (f_M{}^*/f_A f_B) \exp - (\varDelta \varepsilon^*/kT),$$

where $\varDelta \varepsilon^*$, the *potential barrier*, represents the difference between the energy of the ground state of M and the sum of the ground state energies of A and B:

$$\varDelta \varepsilon^* = \varepsilon_M{}^* - (\varepsilon_A + \varepsilon_B).$$

Subdividing the energy in the same way as it was done for the equilibria, we obtain

$$k = (kT/h) \, f \exp - (\varDelta \varepsilon_v^* + \varDelta \varepsilon_1^* + \varDelta \varepsilon_d^* + \varDelta \varepsilon_{n1}^* + \varDelta \varepsilon_s^* (T))/kT .$$

This is a simplified though sufficient version of the more complete expression given in reference [84], to which book the reader is referred for a more profound treatment of the problems of this chapter.

MOLECULES REACTING IN THE GROUND STATE; THERMAL REACTIONS

The formula of the previous section will first be applied to reactions in which only molecules in their ground states participate. As we did for equilibria, we will confine the discussion to making comparisons between rates of similar reactions, such as the reaction rates of reactions of one given reactant with various atoms of a given molecule. In this area the static indices give good results. For instance, in the case of alternant hydrocarbons it is observed that the majority of substitution reactions take place at the carbon with the largest index of free valence, irrespective of the nature of the reaction partner. In the case of substituted conjugated molecules or heteroatomic molecules, the situation is more complicated, but good indications may be obtained by making the assumption that radicals prefer to substitute at the carbon with the strongest free valence, electrophilic reactants at the carbon with the biggest charge q, and nucleophilic reactants at the carbon with the smallest charge q. The static indices characterizing de-localized bonds form an excellent bridge between the structure of the bonds and the chemical reactivity. In order to enter the quantitative domain, it is necessary first of all to have an idea of the reaction path, i.e. to contrive a certain structure of the transition state M^*. Those states are actually too short-lived for an experiment to supply significant data. Two models have been proposed for aromatic molecules. Consider the substitution of a reactant Y in benzene. According to Wheland, the intermediate complex should be

In other words, the attacked carbon becomes saturated and the de-localized bond is interrupted, thereby no longer passing close by the atom's core. The de-localization is decreased. The contribution $\Delta \varepsilon_d^*$ of the localized bond to the potential barrier in this model is called the localization energy. However, Evans suggests to write the intermediate complex

This would mean extension of the de-localization. So far, Wheland's model has found the greatest recognition. With the aid of the latter Daudel et al. [60] could demonstrate the existence of a linearly decreasing relationship between the free-valence index and the localization energy for substitutions affecting a carbon atom of an alternant hydrocarbon. A quantity such as the localization energy, by being dependent on the structure of M, accounts to some extent for the reaction path and is often called a *dynamic index*. The preceding relation was the first proof of the existence of quantitative links between static and dynamic indices, between structure and reactivity. These relations are not always very precise, and on the whole Wheland's dynamic indices give slightly better results than the static indices. Dewar and Warford gave an interesting example of the application of the de-localization energy. Table VI gives the percentages of isomers obtained by Schmidt and Heinle for the nitration of phenantrene. Also in this table the values of the localization energy will be found as calculated by a very simple procedure which takes the integral

TABLE VI

Position of attacked atom	$\Delta\varepsilon^*_d$	Percentage of isomer according to	
		Schmidt and Heinke	Dewar and Warford
10	1.80	60	34
1	1.96	0	27
3	2.04	2	25
2	2.18	20	4

β, associated with a C—C-fragment of the localized bond, equal to 1. Contrary to a justifiable expectation the percentages of isomer (and so the rates of the reactions leading to their formation) do not smoothly decrease with increasing $\Delta\varepsilon^*_d$. One might believe that the participation of other terms in the potential barrier could be the reason for this. However, Dewar and Warford [61] believed the divergence to be attributable to experimental errors. So they repeated and brought up to date the experiments performed by Schmidt and Heinle. Their new percentages of isomers have also been entered into the table, and this time we see the percentages of isomers decrease in proportion with the increase of the contribution $\Delta\varepsilon^*_d$ to the potential barrier. This is a fine example of the possibility that quantumchemistry may reveal experimental errors. There exist also cases that Wheland's model fails to perform well. The localization energies do not explicitly take into account the reactant staying outside of the de-localized bond. Wheland's model has to be abandonned when for a certain class of reactants (nucleophilic, for instance) the position of the most reactive center in a molecule depends on the very reactant. An example of such a case is the actions of the ions OH$^-$ and CN$^-$ on salts of 1-alkyl-chinoline. The OH$^-$ goes to position 2 and CN$^-$ to position 4. Bertran [62a, b] et al. showed that Evans' model should have been selected. This model intro-duces the reactant into the de-localized bond; the contribution $\Delta\varepsilon^*_d$ will then directly depend on the parameters representing the reactant, and calculations have shown that for the majority of reactions examined in that light, the values of $\Delta\varepsilon^*_d$ allow of a very good estimate of the relative rate constants.

PHOTOCHEMICAL REACTIONS

The study of photochemical reactions developed parallel with the study of thermo-chemical reactions. By 1946 the first attempts were made at relating the structure of conjugated molecules in their first excited electronic state to their reactivity. A study of butadiene by means of the mesomeric method will be cited as the only example here [63]. In this method functions are associated with classical formulae as used by chemists to represent a same molecule, and the approximate wave function is written as a linear combination of such functions. When the method is exploited to its full potential, the same wave function will be obtained as with the molecular orbital method. For butadiene the ground state will be represented by the function

$$\Psi = a \, \mathit{/\!\!\!\!\wedge} + b \, \mathit{\triangle\!\!\!\!\!\!\!\!\!\!\_}$$

where the formulae actually symbolize the functions associated with these formulae by the method of mesomers. A calculation gives that $|a| \gg |b|$. This result means that the ground state of butadiene is fairly well represented by the formula $\mathit{/\!\!\!\!\wedge}$. In other words, the extreme fragments of the de-localized bond have a high value of the bond index whereas those of the central fragment have a low value.

For the first excited state we may write

$$\Psi^{\displaystyle *} = a' \, \mathit{/\!\!\!\!\wedge} + b' \, \mathit{\triangle\!\!\!\!\!\!\!\!\!\!\_}$$

and orthogonality requires that $aa' + bb' = 0$, neglecting the overlap integral between the two formulae-functions. We have then $|b'/a'| = |a/b|$ and so $|b'| \gg |a'|$. The formula $\mathit{\triangle\!\!\!\!\!\!\!\!\!\!\_}$ is a rather good one for the excited state; the central fragment has a high value of its bonding index and the extreme ones a low value. Positions 1 and 4 have a high freevalency index. They must be very reactive.

The notation for the excitation from the ground state of a molecule of butadiene by a photon, is

$$\mathit{/\!\!\!\!\wedge} + h\nu \rightarrow \mathit{\triangle\!\!\!\!\!\!\!\!\!\!\_}$$

and suggests the possibility of a photochemical cyclization under formation of cyclo-butene:

$$\mathit{/\!\!\!\!\wedge} + h\nu \rightarrow \mathit{\triangle\!\!\!\!\!\!\!\!\!\!\_} \rightarrow \square .$$

When the results of these calculations were published in 1946, there were no experimental data available regarding the chemical reactivity of the first excited electronic state of butadiene. Then, in 1963, Srinivasan [64] obtained cyclobutene by irradiation of butadiene dissolved in ether. This provided proof of the heuristic value of quantumchemistry. Attempts were made to employ Wheland's static- and the dynamic indices for an interpretation of photochemical reactions [65a, b]. It was found that Wheland's indices never led to good results. The static indices were shown to give superior results. The cause underlying this observation is easy to understand. In photochemical reactions there are no or only very small potential barriers. Necessarily then, the transition state is very similar to the initial state of the molecule attacked, and so the static indices are of great importance. Wheland's model requires a thorough modification of the de-localized system, and such a situation will never be reached in the course of a photochemical reaction. Evans' model, on the other hand, appears to be quite suitable for photochemical purposes. On the basis of this model it was recently possible to explain the orienting effect of substituents on the photo-oxidation reactions of a series of acenes [66]; Wheland's model and even the static indices, gave in that case results diametrically opposite to the experimental findings.

THE CONTRIBUTION OF QUANTUMCHEMISTRY TO CHEMICAL CARCINOGENESIS

Cancer is an illness, or rather a group of illnesses, characterized by an anarchical development of so-called cancerous cells. The pathological character acquired by one cell is subject to transmission in the course of the cell division. In other words, the cells generated by the division of a cancerous cell are (very often) cancerous themselves. Cancer is said to be hereditary at the cellular level. We are tempted therefore, to interpret a cancerous condition as the result of a mutation (or at least modification) of the genetic code, i.e. of the unmasked portions of the DNA enclosed in the cell. Numerous chemical substances are known to be capable of provoking cancerous tumors. The area of the aetiology of cancer devoted to the study of the formation mechanisms of tumors under the influence of chemical substances, is the domain of chemical carcinogenesis. From a teleological point of view this discipline envisages the improvement of the prophylaxis and the therapy of cancer. Two big collections of hypotheses dominate the study of chemical carcinogenesis. According to one, the malignant substances intervene directly on the level of the nucleic acids, thus producing a veritable mutation. According to the other, the action of carcinogenic substances being with the alteration of certain enzymes and consequently with a fixation on proteineic material. It is conceivable that the two mechanisms occur simultaneously, in a relative order of importance as a function of the carcinogenic substance studied.

The family of the conjugated molecules comprises several carcinogenic substances, and it is a striking observation that very slight structural differences may mean considerable differences in carcinogenic potential. Quantumchemistry facilitated the establishment of relationships between the electronic structure of the de-localized electronic bond system and the carcinogenic potential. We will briefly summarize some results regarding alternant hydrocarbons [67a, b].

Chemically speaking, the alternant hydrocarbons can participate in two distinct types of addition reactions. Some reactions, such as the diene synthesis, affect the hydrocarbon in two sites para-oriented with respect to each other:

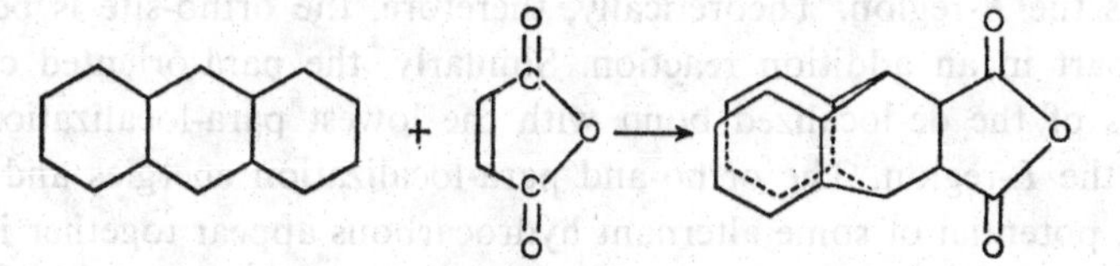

Other reactions take place on adjacent carbons:

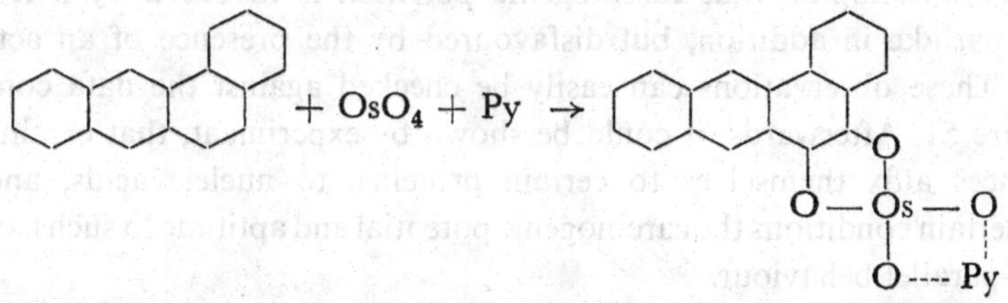

However, in the early stages of the search for links between chemical reactivity and carcinogenesis, there was a total lack of data regarding the relative rates of such reactions within a family of similar molecules. Even at the present time several gaps persist. It was necessary to resort to theoretical estimates. Wheland's model can be generalized in the case of additions. The significant transition state basically resembles one that would result for a reaction of either the first or the second type, from two simultaneous substitutions at two para- or ortho-oriented positions respectively. The contribution $\Delta\varepsilon_d^*$ of the de-localized

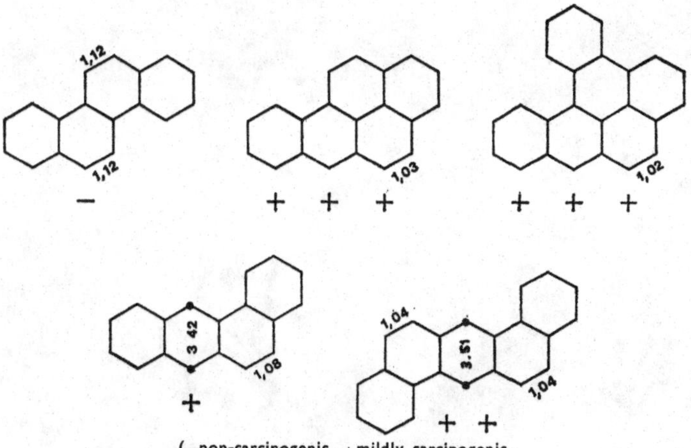

(− non-carcinogenic, + mildly carcinogenic,
+ + medium carcinogenic, + + + very carcinogenic)

Fig. 51.

bond to the potential barrier is called the para-localization energy or ortho-localization energy, respectively. The fragment of the de-localized bond which for a given molecule possesses the lowest ortho-localization energy, has become known as the K-region. Theoretically, therefore, the ortho-site is best-suited to take part in an addition reaction. Similarly, the para-oriented couple of fragments of the de-localized bond with the lowest para-localization energy is called the L-region. The ortho-and para-localization energies and the carcinogenic potential of some alternant hydrocarbons appear together in Figure 51. A Pullman (see references in works cited) showed on the basis of a large number of examples, that carcinogenic potential is favoured by a K-region apt to partake in addition, but disfavoured by the presence of an active L-center. These observations can easily be checked against the data contained in Figure 51. Afterwards, it could be shown by experiment, that carcinogenic substances affix themselves to certain proteins, to nucleic acids, and that under certain conditions the carcinogenic potential and aptitude to such fixations, exhibit parallel behaviour.

CONCLUSION

Epistemological Considerations Regarding the Quantum Theory of the Chemical Bond

Towards an Epistemology of Quantum Theories

We will conclude this book by giving some attention to the epistemology [68] of quantumchemistry, the meeting place of physics and chemistry. In my opinion Bachelard [3] accomplished the complete separation of its fundamental elements from the heuristic climate so typical for this kind of science, born out of dialectics confined to applied rationalism and technical materialism. In this area of science experiments designed 'to see' have become rare, indeed. The researchers perform their experiments in order to extract from them data to denounce or corroborate a theory, in order to compare quantities between which they presume certain correlations. In other words, the volume of rational structures in their minds predispose them to organize one experiment rather than another, to scrutinize one fact rather than another. In their world the intellectual evolution takes place by closely following the appearance of new experimental results. The introduction of concepts and the formulation of new theories aim at the immediate integration of the newly acquired empirical knowledge, the prediction of new phenomena and the experimental arrangements to unveil these. In the words of Bachelard, scientific knowledge not only harpoons reality, it anchors on its grounds. The far-reaching specialization required to perform such tasks often compels the researcher to one of the functions of theoretician or experimenter. The initial dialectic tends to bifurcate: the most theoretical part of the experimenter's thinking provides the basis for the theoretician's meditations, whereas the latter's achievements gradually modify the experimenter's rational structures and language. As the co-founder with Professor Rumpf of a group uniting experimenters and theoreticians, which has been uninterruptedly active since 1950, I have been able to watch closely the formative mechanism of such exchanges. Quantumchemistry is an application of quantummechanical methods, and therefore its epistemology must contain many of the elements that characterize quantummechanics. It should be born in mind that the attribution of a mathematical operator acting on the wave function, to each quantity open to measurement, is its first principle. And let us recall that a measurable quantity essentially amounts to a set of operations leading to a number, called measure. An operator symbolizes a sequence of rational operations. When we consider a concrete system characterized by an ensemble of quantities, that first principle establishes an application of the ensemble of quantities into an ensemble of operators, an application of an ensemble of ensembles of material operations into an ensemble of ensembles of rational operations; the possible values to be attached to each quantity coincide with the eigenvalues of the associated operator. A closer analysis of the first principle also brings us to the recognition that a quantity which is the sum or the product of other quantities, must be associated with an operator which in turn is the sum or the product of the associated operators. The ensemble of quantities and the ensemble of associated operators have been supplied with internal laws and the first principle establishes an isomorphism between the two structures: that of the material operations and that of the rational operations. The logical nature of quantum-reasoning has above all to be structuralistic. More of such structuralistic elements may be discovered. Xenakis accentuated the isomorphism between musical objects, algebraic entities and geometrical objects, and taking his departure at this remark was able to unify architecture, music and mathematics.

The second principle defines the equation to be obeyed by a wave that conducts the movements of the particles in a system.

The third principle permits us with the aid of that wave, to calculate the probability of obtaining a particular result from the measurement of a characteristic quantity of the system. In the majority of actual cases that probability will be less than one. Certainty will never be attained. The wave Ψ in wave mechanics is a probability wave. Wave mechanics contains the dualism between necessity and contingency since, although an event may be possible, it is not open to a rigourous calculation of its probability. This is the expression of a subdued determinism in remarkable conformity with Democrites' assertions leading to his conclusion that all things in the Universe are the fruit of chance and necessity. It is interesting to underline the special nature of the probability-information pertaining to one individual only. Quantum mechanics allows us to calculate the average lifetime or half-life of a radioactive atom. An experiment will provide its magnitude. We know, for instance, that the half-life of beryllium-7 is 52 days. We can deduce that of each gram of this metal prepared to-day, only one half gram will be left 52 days from now, the other half gram having become lithium. The probability-information about average lifetime informs us precisely on the behaviour of a large population of particles, because probabilities pertaining to large populations transform into certainties. However, should we be interested in the fate of one single radioactive nucleus, its average lifetime tells us strictly nothing. It may disintegrate immediately or by the end of eternity. So everything is contingency; Jacques Rueff [69] drew the conclusion that particles enjoy a certain freedom [70], that determinism only applies to the macroscopic world, to vast collections of individuals but not to things on the level of particles and individuals. Although this ingenious thesis cannot be scientifically contradicted, one should remain aware that it is based on a hypothesis. It assumes the necessity of the stochastic character of quantum-determinism. Louis de Broglie [71] has technically demonstrated that this hypothesis is not self-imposing but that the stochastic character could be the result of our own ignorance of hidden parameters, in which case its origin has to be found in the may be transient, experimental deficiencies. (de Broglie has shown that the wave Ψ undoubtedly conceals a more 'material' wave u.) Finally, I believe that attention should be drawn to the strangely subjective character of wave mechanics. Let us view the evolution of a system of two electrons with parallel spins. When we want to find electron 1 in point M_1 and electron 2 in point M_2, the wave function for this expectation is $\Psi(M_1, M_2)$. On the other hand, should we look for electron 1 in M_2 and vice versa, the wave function would be $\Psi(M_2, M_1)$. By virtue of the Pauli principle we must have $\Psi(M_1, M_2) = -\Psi(M_2, M_1)$, and the wave changes sign upon modification of our expectation; this sign-change is merely operational. In fact, when we try to find both electrons at M_1, we must put $M_1 = M_2 = M$, and we would obtain

$$\Psi(M, M) = -\Psi(M, M) = 0.$$

The change of sign implies the impossibility to find two electrons with parallel spins in one small volume of space. It could seem that this fact sheds light on the nature of the intellectually-conceived entities occurring in the isomorphisms inserted by quantummechanics between the rational and the material, between the true and the real. At any moment, the wave associated with two points in space where the corpuscle may be found, depends on the nature of the question one asks in this respect.

On the Epistemology of the Chemical Bond

In our opinion the theory of loges is the greatest advancement of applied rationalism towards the integration of the notion of a bond into a logico-mathematical structure. We noticed that the concept of loge arose from an apparent contradiction between the languages of chemists and physicists. The theory of loges had its teleological origin in the desire to establish a synthesis between two opposites. In order to account for the existence of chemical bonds via an electronic theory, chemists followed Lewis in making a distinction between valence electrons and core electrons, and they proposed an association between certain valence electrons and a given bond, between other valence electrons and other bonds, they drew distinctions between K-, L-, σ- and π-electrons. The electrons were labeled, individualized, localized. Strong emphasis was laid on the corpuscular aspects.

The thoughts of the theoretical physicists, on the other hand, had a tendency to focus on wave equation and wave function, and on the fundamental principles: the indistinguishability of the electrons and the Pauli principle imposing severe restrictions on the symmetry of the wave. But how to attribute different names to a molecule's electrons when the latter are indistinguishable? How can we have them play such distinct roles when the symmetry of the wave function requires that they be equal by virtue of some egalitarian principle? How to localize electrons along particular bonds when the wave exists only to the extent that they are not localized? On one hand, there is the full weight of the empirical way of thinking based on a multitude of observations, and on the other hand the logical requirements of principles brought about in an attempt at a synthesis possible only when that multitude of empirical facts is totally neglected? This is the 'between parentheses', the quasi-Husserlian exposition mentioned by Suzanne Bachelard [72]. Should one try to give a deductive account of knowledge acquired previously through contacts with the experiment, one should forget about it.

The structuralism we talked about, constituted a kind of bridge between 'existence' and 'essence'. Descending from the principles to a line of thought emerging out of the desire to understand the real, it is not always easy to find the passage; the theory of loges is a road beginning at the principles of quantum-mechanics and leading back to chemical intuition, adding far more rigourous features to this latter. Jørgensen called it an 'aristic' theory resulting from reflections on the shape of the wave associated with a molecule. At any moment, this wave has a significant value in a very small space-element D only. Then there is the quasi-certainty that all electrons and nuclei may be found in that element. Strong correlations also exist between the movements of the nuclei: the distances between two points with a high probability of the presence of two nuclei A and B, are confined to a small interval. In other words, the wave function contains a certain nuclear topology. The infinitely rich information conferred on the electrons by the wave function, is far more diffuse and complex, though, and it is necessary to transform part of this information into a limited and yet evocative set of numbers. The theory of loges for a molecule with n electrons, divides space into p unconnected volume elements, together still covering all space and with p confined to the open interval 1, n. Subsequently, one calculates the number of ways in which the electrons can be distributed over these volume elements. Information theory provides here the facilities to measure the amount of information conferred on the molecule by each partition of the space. For each value of p the amount of information in every volume element is determined as a function of the limits chosen for each element. The maximum in the amount of information determines the best division into p loges. This sequence is repeated for all values of p. The ultimate, maximum information characterizes the best division into loges. As is usually the case in quantummechanics, the complete execution of that sequence of operations is practically impossible and we must confine ourselves to an approximation. In simple cases the approximation certainly remains profoundly meaningful; very often it happens that the chemist draws a line between two atoms, indicating for instance two localized electrons, and that the best division into loges includes, between the quasi-spherical volumes, the so-called core loges surrounding the nuclei of the atoms in question, a volume (abutting upon the core loges) where the probability of finding two and only two electrons, is high; this is the so-called bonding loge. With two electrons thus marked and localized corresponds a well-defined volume in space where a high probability prevails that two and only two electrons may be found there. In this way the theory of loges contributes to the concept of the chemical bond a notion about space, necessity and contingency. The necessity resides in the fact that a repetition of the measurement of the number of electrons in a loge will very often give the result two. The contingency means that in a particular measurement any number between 1 and n may be found in the loge, but that when two are found they may be any two. Bonds shall not be associated with particular electrons but with volume elements in the molecular space, visited in some particular manner by the ensemble of electrons.

The theory of loges illustrates rather strikingly the way in which quantummechanics represents two antagonistic dynamics (in the sense as defined by Lupasco [73]) in a case that their potentialization and actualization plays equally important roles. According to Wolff [74] the value of a doctrine 'is not so much its actual value, which actually cannot be established, but rather its heuristic value'. No matter how new the theory of loges may be, its heuristic power has made itself felt already. This heuristic power finds its roots undoubtedly in the rigour of the definition of the best division into loges, in the clarity and simplicity of this concept, and in the fact that it favours one partition into fragments of the molecular space. This fragmentation of the molecular space suggests the fragmentation of molecular quantities, and thereby justifies the systematics of additive molecular properties as these were empirically discovered by the experimenters. In some cases it may be necessary to add terms to the sum of moduli associated with the loges accounting for the interaction between adjacent loges, and on their basis it was possible to derive from the theory of loges a simple method for the calculation of isomerization energies. The concept's clarity makes it possible to clearly distinguish between covalence and dative bond on indisputable grounds and no longer on the basis of merely plausible and often purely conventional mechanistic data. Eventually, the idea of the loge has been used as the starting point for a new method of calculating wave functions. Far less delicate to handle than the notion of orbital, less easily subject to the suggestion of false notions, more directly linked to chemical intuition, the idea of the loge seems to have a greater pedagogical value and should, in our opinion, be taught before the more technical orbital.

Although our reflections on quantum chemistry were doomed to remain very superficial, we hope to have made tangible that in this domain scientific knowledge constitutes a 'reality', an intimate union of observation, action and thinking, of concrete and abstract elements. The goals of epistemology are to examine how knowledge takes shape, to analyze the nature of this absolute relative knowledge, to reveal its logic origin, its value, the range of its objectivity. We have penetrated to its frontiers.

REFERENCES

[1]   L. de Broglie, Dissertation, Masson, Paris, 1924.
[2a] Ø. Burrau, Det. Kgl. Danske Vid. Selskab. 7, 1 (1927).
[2b] W. Heitler and F. London, Z. Phys. 44, 455 (1927).
[3]   G. Bachelard, Le Rationalisme Appliqué, Presses Universitaires de France, Paris, 1966.
[4]   P. W. Bridgman, The Logic of Modern Physics, Macmillan, London, 1928.
[5]   Yukawa and Sakata, Proc. Phys. Math. Soc. Japan 17, 467 (1935).
[6]   L. W. Alvarez, Phys. Rev. 52, 134 (1937).
[7]   E. A. Hylleraas, Z. Phys. 65, 209 (1930).
[8]   T. Kato, Trans. Am. Math. Soc. 70, (2) 212 (1951).
[9a] R. Daudel, Compt. Rend. Acad. Sci. 237, 60 (1953).
[9b] R. Daudel, S. Odiot, and H. Brion. J. Chim. Phys. 51, 74 (1954).
[9c] H. Brion, R. Daudel, and S. Odiot J. Chim. Phys. 51, 358 (1954).
[9d] S. Odiot and R. Daudel, J. Chim. Phys. 51, 361 (1954).
[10] H. M. James and A. S. Coolidge, J. Chem. Phys. 1, 825 (1933).
[11] W. Kolos and C. C. J. Roothaan, Rev. Mod. Phys. 32, 219 (1960).
[12] M. Roux, S. Besnainou, and R. Daudel, J. Chim. Phys. 53, 218 (1956).
[13] S. Odiot, Compt. Rend. Acad. Sci. 237, 1399 (1953).
[14] S. Odiot and R. Daudel, Compt. Rend. Acad. Sci. 238, 1384 (1954).
[15] J. W. Linnett and A. J. Poe, Trans. Faraday Soc. 47, 1033 (1951).
[16] J. Dalton, New System of Chemical Philosophy (1808, 1817).
[17] G. N. Lewis, J. Am. Chem. Soc. 38, 762 (1916).
[18a] T. M. Lowry, Trans. Faraday Soc. 18, 285 (1923).
[18b] T. M. Lowry, J. Chem. Soc. 1923, 822 (1923).
[18c] G. N. Lewis, Valence and the Structure of Atoms and Molecules, Chemical Catalog Co., New York, 1923.
[18d] N. V. Sidgwick, The Electronic Theory of Valency, Oxford University Press, Oxford, 1927.
[19] R. Daudel, Structure Electronique des Molecules, Gauthier-Villars, Paris, 1956.
[20] H. Brion, R. Daudel, and S. Odiot, J. Chim. Phys. 51, 553 (1954).
[21] C. Aslangul, Compt. Rend. Acad. Sci. 272, 1 (1971).
[22] K. Jørgensen, Nature et Propriétés des Liaisons de Coordination, C.N.R.S., p. 21 (1970).
[23a] M. Roux, S. Besnainou, et R. Daudel, J. Chim. Phys. 218, 939 (1956).
[23b] M. Roux, J. Chim. Phys. 220, 754 (1958).
[24] S. Bratos, R. Daudel, M. Roux, and M. Allavena, Rev. Mod. Phys. 32, 412 (1960).
[25] P. Pascal, Ann. Chim. Phys. 19, 70 (1910).
[26] R. Daudel and F. Gallais, Rev. Chim. Min. 6 61 (1969) and references therein.
[27a] R. Daudel, Les Fondements de la Chimie Théorique, Gauthier-Villars, Paris 1956; English translation: The Fundamentals of Theoretical Chemistry, Pergamon, Oxford, 1968.
[27b] R. Daudel, F. Gallais, and P. Smet, Int. J. Quant. Chem. 1, 873 (1967).
[28] R. Daudel and A. Veillard in Nature et Propriétés des Liaisons de Coordination, C.N.R.S., 1970, p. 15.
[29] P. O. Löwdin, J. Mol. Spectrosc. 3, 46 (1959).
[30] R. Daudel, Theorie Quantique de la Réactivité Chimique, Gauthier-Villars, Paris, 1967;

English translation: *Quantum Theory of Chemical Reactivity*, D. Reidel Publishing Company, Dordrecht, 1973.

[31] J. Lennard Jones and J.Pople, *Proc. Roy. Soc.* **202**, 166 (1950).

[32] H. Brion and R. Daudel, *Compt. Rend. Acad. Sci.* **237**, 567 (1953).

[33] J. M. Forster and S. F. Boys, *Rev. Mod. Phys.* **32**, 300 (1960).

[34] G. Berthier in *Aspects de la Chimie Quantique Contemporaine*, C.N.R.S., 1971, p.49.

[35] R. Lefebvre and Y. Smeyers, *Int. J. Quant. Chem.* **1** 403, (1967).

[36] J. Pople in *Aspects de la Chimie Quantique Contemporaine*, C.N.R.S., 1971, p. 17.

[37a] M. Klessinger and R. McWeeny, *J. Chem. Phys.* **42**, 334 (1965).

[37b] R. McWeeny, *Proc. Roy. Soc.* **A253**, 242 (1959).

[38] E. Hückel, *Z. Phys.* **70**, 204 (1931).

[39] C. Sandorfy and R. Daudel, *Compt. Rend. Acad. Sci.* **238**, 93 (1954).

[40] R. Hoffmann, *J. Chem. Phys.* **39**, 1397 (1963).

[41] R. Pariser and R. G. Parr, *J. Chem. Phys.* **21**, 466 and 767 (1953).

[42] J. A. Pople, D. P. Santry, and C. A. Segal, *J. Chem. Phys.* **43**, 129 (1965).

[43a] R. Daudel, R. Lefebvre, and C. Moser, *Introduction to Quantum Chemistry*, Interscience, New York, 1959.

[43b] R. Daudel and C. Sandorfy, *Empirical and Semi-Empirical Wave Mechanical Calculations*, Yale University Press, New Haven and London, 1971.

[44a] L. Pauling, *Proc. Nat. Acad. Sci.* **18**, 293 (1932).

[44b] C. A. Coulson, *Proc. Roy. Soc.* **169**, 413 (1939).

[45] C. A. Coulson and G. B. Rushbrooke, *Proc. Camb. Phil. Soc.* **36**, 193 (1940).

[46] C. Vroelant and R. Daudel, *Bull. Soc. Chim. Fr.* **16**, 36 (1949).

[47a] R. Daudel and A. Pullman, *Compt. Rend. Acad. Sci.* **220**, 888 (1945).

[47b] Svartholm, *Arkiv. Kemi Mineral Geol.* **15A** (13)...(1941).

[47c] C. A. Coulson, *Trans. Faraday Soc.* **42**, 106, 265 (1946).

[48] R. Daudel and A. Pullman, *Compt. Rend. Acad. Sci.* **222**, 663 (1946).

[49] O. Chalvet, R. Daudel, and F. Peradejordi, *J. Chim. Phys.* **59**, 703 (1962).

[50] G. J. Hoytink, E. de Boer, P. H. van der Mey, and W. P. Weijland, *Rec. Trav. Chim.* **75**, 487 (1956).

[51a] J. Bertran, O. Chalvet, and R. Daudel, *Theor. Chim. Acta* **14**, 1 (1969).

[51b] J. Bertran, J. Dannenberg, R. Leute, C. Ponce, O. Chalvet, and R. Daudel, *Theor. Chim. Acta* **17**, 249 (1970).

[51c] J. Bertran, O. Chalvet, and R. Daudel, *Ann. Fisica* **66**, 123 (1970).

[52] J. Monod, *Le Hasard et la Nécessité*, Editions du Seuil, Paris, 1970.

[53] C. Ponnamperuma and N. W. Gabel, *Space Life Sci.* **1**, 64 (1968).

[54] S. L. Miller, *Science* **117**, 528 (1953).

[55a] C. Ponnamperuma, C. Sagan, and R. Mariner, *Nature* **199**, 22 (1963).

[55b] C. Ponnamperuma, R. M. Lemmon, R. Mariner, and M. Calvin, *Proc. Nat. Acad. Sci.* **49**, 737 (1963).

[56a] B. Pullman, *La Biochimie Electronique*, Presses Universitaires de France, Paris, 1969.

[56b] A. Pullman and B. Pullman, *Quantum Biochemistry*, Interscience, New York, 1963.

[57] A. Pullman, *Electronic Aspects of Biochemistry*, Academic Press, New York, 1964.

[58] L. B. Kier in *Fundamental Concepts in Drug Receptor Interactions*, Academic Press, New York, 1970.

[59] F. Peradejordi in *Aspects de la Chimie Quantique*, C.N.R.S., 1971, p. 321.

[60] R. Daudel, C. Sandorfy, C. Vroelant, P. Yvan, and O. Chalvet, *Bull. Soc. Chim. Fr.* **17**, 66 (1950).

[61] M. J. S. Dewar and E. W. T. Warford, *J. Chem. Soc.* **1956**, 3570 (1956).

[62a] J. Bertran, O. Chalvet, R. Daudel, T. F. W. McKillop, and G. H. Schmid, *Tetrahedron* **26**, 339 (1970).

[62b] O. Chalvet, R. Daudel, and T. F. W. McKillop, *Tetrahedron* **26**, 349 (1970).

[63] A. Pullman and R. Daudel, *Compt. Rend. Acad. Sci.* **222**, 288 (1946).

[64] R. Srinivasan, *J. Am. Chem. Soc.* **85**, 4045 (1963).

[65a] R. Srinivasan, *Reactivity of the Photo-excited Organic Molecule*, John Wiley, New York (1965).

[65b] R. Daudel in *Advances in Quantum Chemistry*, Vol. 5, Academic Press, New York, 1970.

[66] O. Chalvet, R. Daudel, G. H. Schmid, and J. Rigaudy, *Tetrahedron* **26**, 365 (1970).

[67a] A. Pullman and B. Pullman, *Cancérisation par les Substances Chimiques et Structure Moléculaire*, Masson, Paris, 1955.

[67b] P. Daudel and R. Daudel, *Chemical Carcinogenesis*, John Wiley, New York, 1966.

[68] A. Lalande, *Vocabulaire Technique et Critique de la Philosophie*, Presses Universitaires de France, Paris, 1947.

[69] J. Rueff, *Les Dieux et les Rois*, Hachette, Paris, 1968.

[70] R. Ruyer, *La Genèse des Formes Vivantes*, Flammarion, Paris, 1958.

[71] L. de Broglie, *La Physique Quantique Restera-t-elle Indéterministe?*, Gauthier-Villars, Paris, 1953.

[72] S. Bachelard, *La Conscience de Rationalité*, Presses Universitaires de France, Paris, 1959.

[73] S. Lupasco, *Les Trois Matières*, Julliard, Paris, 1960.

[74] E. Wolff, *Les Chemins de la Vie*, Hermann, Paris, 1965.

INDEX OF SUBJECTS